The Unspeakable Truth

How Fake History, Christianity, and Self-Oppression Robs African Americans of Their Voice in Modern UFO Revelations

Roderick Martin

4BiddenKnowledge

ISBN 979-8-9871224-9-5

Books may be purchased by contacting the publisher and author at:

https://www.4biddenknowledge.com/online-store

Publisher info:

4biddenknowledge Inc

http://4BK.TV

4biddenknowledge.com

Info@4biddenknowledge.com

FOREWORD

By Billy Carson, Founder & CEO of 4biddenknowledge Inc.

There are moments in human history when truth can no longer remain buried, when the pressure of suppressed knowledge becomes so great that it must rise to the surface. We are living in one of those moments now. For thousands of years, humanity has been fed a carefully edited version of its own origins, one shaped by power, religion, colonization, and control. *The Unspeakable Truth* is not just a book that challenges history, it confronts the very foundation upon which modern civilization has been built.

When I first encountered the work of Roderick Martin, I immediately recognized something rare: a researcher willing to follow the evidence wherever it leads, even when it collides with deeply held beliefs, institutional narratives, and cultural conditioning. This is not easy work. It requires intellectual courage, spiritual resilience, and a deep respect for truth over comfort. Roderick possesses all three.

This book boldly explores the intersections of ancient civilizations, extraterrestrial contact, biblical reinterpretation, lost global cataclysms, suppressed archaeology, and genetic anomalies, but it does so through a lens that has been historically ignored: the African American and

Indigenous perspective. For far too long, these communities have been excluded from the broader conversation around UFOs, ancient knowledge, and humanity's true origins, despite the rich ancestral records that exist within their cultures. This book restores that missing voice.

The Unspeakable Truth forces us to ask uncomfortable but necessary questions:

• What if the gods of our ancient texts were not symbolic, but literal visitors from the stars?

• What if the Great Flood was not a myth, but a global reset that erased an advanced civilization?

• What if entire chapters of human history were deliberately rewritten to suppress remembrance of who we truly are?

• And what if modern religion itself is a repackaged memory of extraterrestrial contact?

These are not speculative questions meant to entertain. They are questions supported by star-aligned megaliths, anomalous skulls, forbidden archaeological sites, ancient DNA, and worldwide creation narratives that all tell the same story using different names. When thousands of independent cultures across the planet describe the same celestial beings, catastrophic floods, giants, and sky gods, coincidence is no longer a sufficient explanation.

What makes this work especially powerful is that it is not written from a place of academic detachment. Roderick shares his personal awakening, from a traditional Southern Christian upbringing to becoming a MUFON investigator and truth researcher. His journey mirrors the very transformation now happening across the world as governments acknowledge what many of us have long known: we are not alone, and we never have been.

As the founder of 4biddenknowledge, I have spent years uncovering the connections between ancient texts, advanced technology, lost civilizations, and extraterrestrial contact. This book aligns with that mission perfectly. It adds a critical dimension to the global awakening, one that reconnects African Americans to a suppressed cosmic legacy and dismantles the psychological chains of imposed history and inherited limitation.

This is not a book for the complacent.

This is a book for the seeker, the awakened, the rebel thinker, and the soul who senses that the official story has never been complete.

If you choose to continue reading, do so with an open mind and a courageous heart. You are about to encounter ideas that may disrupt what you thought you knew about religion, history, race, and humanity's place in the universe. But disruption is often the first step toward liberation.

Truth does not need permission to exist.

It only needs someone brave enough to speak it.

And now, it has been spoken.

—Billy Carson

Founder & CEO, 4biddenknowledge Inc.

Researcher, Author, Truth Seeker

Works by 4BIDDENKNOWLEDGE

Check Out Other Books by Billy Carson

Listen to the Music by Billy Carson

Contents

Part Three
The Plot Twist of Biblical Proportions

Preface

I grew up in the era of "It takes a village to raise a child," and my street was one large community. There was a church on nearly every corner in a repurposed house. Everyone attended Sunday school and church on Sunday afternoon and Wednesday night. All the families on the street were black, and everyone lived on food stamps and Jesus. It was the Bible Belt in the 1970s, and while life was hard in the black community, our blue-eyed and brown-haired Jesus promised nothing but struggles during life with heaven everlasting after death. As a single parent, my mother was utterly devoted to the Baptist church and saw God as her actual father due to the absence of her birth father. Over time, she unconsciously pushed this same belief onto her children.

After my parents divorced, I moved back and forth between both homes, making it difficult to make friends. In my small community, all the kids attended the same school and lived in the same neighborhood. We were all taught the same lessons by our parents: survival is the only thing we are able to do, always be on guard for racism or racial segregation, and always "stay in your lane" when maneuvering through life. After school, all the neighborhood kids and I would have our bonding time outside until the sun came down. Our moms would flick the porch light on and off, signaling it was time to come in. Walking towards the front door, I would look up at the twilight sky and wonder if there was more out there in the vast openness of space.

One night, as I walked inside, I looked up towards the stars and saw a UFO. What should have been a moment suspended in time was a simple awestruck moment amid everyday life. I did not tell anyone about what I saw for fear of ridicule. My nickname when I was younger was Coocoo, after I was fascinated with my grandmother's Coocoo clock, and the last thing I wanted was to be labeled as genuinely crazy. I had seen UFOs in comic books, magazines, and space stories, but they were difficult to comprehend with the teachings of the Bible. I kept the moment in the back of my mind and continued with life; after all, I was in a family of athletes, and in Texas, football is second only to God.

In school, I was taught that Christopher Columbus discovered America and that all black people were brought from Africa to America on slave ships. The only culture I knew was survival in the southern Bible Belt. I was taught that I was oppressed because of my skin color and that I would be fighting for freedom and equality for the rest of my life. It was not until I went off to college at the University of Kansas that I began to transition out of my all-black community. The culture shock of the university started to open my eyes to a different reality. I was one of 500 black students in a school of 30,000. While I was focused on getting through college and going pro in football, other students were attempting suicide because of subpar grades. My 'survival' worldview versus another's worldview of 'education determines elevation' was a shock to my entire upbringing and was the springboard that changed the course of my life.

Growing up, I was taught that God created the earth in five days, created man on the sixth day, and rested. I was taught that man was the pinnacle of creation and that nothing existed outside Earth. Every sermon on Sunday and Wednesday was about how I was a sinner and needed Jesus, and somehow they turned into why I needed to tithe. There was no room for the vastness of the universe or the possibility of other lifeforms out there.

It was only when I was much older that I was able to grow spiritually and understand the stronghold of the mind that Christianity employed. After exploring spirituality, I returned to that night when I was twelve and began diving deeper into UFOs. Despite my wife's request at the time, I became obsessed and have never looked back. I became a Mutual UFO Network (MUFON) investigator, began speaking about UFOs on Clubhouse, and started a YouTube channel called High Strangeness. I began to realize that just like that little Southern Baptist kid who saw a UFO, our eyes are useless when our mind is blind; it is time to ask why.

It was not until I began researching the UFO phenomenon that I began to uncover a thread of myths and mysteries. This thread has been carried throughout generations as a type of underground movement. I became curious about why this movement was so protected and challenging to access. When I began looking at cuneiform tablets, I started to understand that there was a creation story that was more intricate than the Bible, and my ears perked up. I stumbled on the *Atra-Hasis* and learned about the arrival of the Anunnaki to Earth and how they worked the land and created rivers or "separated the waters from another," just like Genesis chapter one describes. The *Atra-Hasis* introduced me to how the Anunnaki gods conspired to create modern man using advanced technology.

The ancient stories tell of extraterrestrials living alongside humanity. Every human race lived with these beings, and their tales echo through history as legends and fables. Today, we experience these echoes as UFO sightings or contact with extraterrestrial beings. As a UFO researcher, I have noticed an interesting trend in the reporting of UFO sightings and stories. After speaking with friends working in this field, the harsh truth comes to light: there are crucial experiences missing in the UFO community because the African and Native American communities do not talk about their experiences. This is fascinating to me because there is such a rich history in these tribes of people. Our stories of contact have been lost to time. Yet, with the discoveries made in the past three centuries, old stories have become new and they reveal a different history than we have been told.

As I began to study various mythological stories all over the globe, I was stunned to realize that while the names change throughout multiple regions and across time, the attributes of these ancient extraterrestrial ancestors stay the same. To bring hope to our future, we need to understand our past. To understand our past, we need to go beyond the story that humanity has been told for 5000 years, look at different myths or legends all over the globe, and piece together a cohesive narrative that reveals a very different story of what is happening today.

I have spent countless meticulous hours sifting, sorting, adding, subtracting, and rewriting to present a comprehensive and deep dive into each subject. The greatest care was taken to create a literary picture that flows. To encapsulate the known and unknown history of humanity is a mystery that still eludes everyone today, and to say otherwise is complete nonsense.

This book looks at nonconventional views alongside Biblical stories to give the reader a different perspective to begin asking questions. All information is sourced, and I invite the reader to look into each source. As this book came to life, it was a journey of discovery, disbelief, and—at times—shocked annoyance at how alternative stories aligned better than the history we are told today. I hope that you will enter with an open mind to uncover *The Unspeakable Truth* that, when spoken aloud, could have the power to transform minds.

When Myths Tell the Truth

Imagine that the year is 5000 Common Era (CE), and humanity has colonized Mars, the moons of Jupiter, and other planetary bodies within our solar system. Collectively, we have learned the art of Genesis sciences and can terraform planets to our will. We have moved beyond the solar system and have reached the Andromeda star system. Upon arrival on a young planet, humanity finds primitive lifeforms that resemble the ones spoken about in the annals of humanity's history; cave people of a new world. This planet contains vast resources that can help build humanity's empire to new heights and continue their galaxy exploration. Other colonies of humans across the galaxies hear about this discovery. Various colonies send select groups of Genesis scientists to this newfound oasis.

As with all missions, there is a leader, command structure, and the workers. The mission of these Genesis scientists' groups is to terraform this planet. After decades of work, the workers become exhausted and groan to their command staff about the great labor. The command staff listens to their cries and works to create helpers by mixing the DNA of one of the scientists with the DNA of a primitive being on this new planet. After trial and error, a hybrid creation is born and ideally suited to the planet's natural habitat. Each group marks their hybrids with a specific genetic marker that causes the skin of their creation to match the skin of their colony. These new versions of primitive lifeforms are similar to us, yet slightly different. A select few scientists from each colony stay and help the hybrids adjust to their surroundings. They teach them agriculture, astronomy, language, mathematics, healing, and many other valuable teachings that can help jumpstart their new population.

The scientists spread the hybrids all around the planet in various groups. Each scientist is placed with a group of hybrids to ensure everything goes according to plan. The scientists build great monuments around the planet and create a planetwide civilization that lives in harmony with one another. The scientists discover a world-ending catastrophe is coming and work to help the hybrids survive. Underground caverns are formed, and ships are built to house the hybrids until the catastrophe ends. Generations pass while waiting for the atmosphere to be inhabitable again. When the hybrids reemerge, the scientists discover that nothing is quite as it used to be, and they need to work to make the planet habitable again.

Lakes are reformed, deserts stand where there were once lush forests, and previous metropolises have vanished under the sea. The scientists work with the hybrids to create new civilizations and pass down teachings of their planet, the sun, the cycles of the stars, and their moon or moons. They give the reemerging hybrids the teachings they once gave their ancestors. Civilization begins again from out of nowhere, or so it seems.

Due to the incredible knowledge that humanity has handed down to these hybrids, the hybrids give them praise and begin to worship them, even going as far as calling them gods. Each human scientist rules over their particular sect of hybrids for thousands of years as both their god and king or queen. Great monuments are erected and artwork is created to revere the "gods" within each group. Over time, the hybrids wish to rule themselves and come to an agreement with the scientists. The various kingships are handed down to the hybrids. The scientists' stories and interactions with them are written down or kept in the cultural memory through oral tradition. These stories are passed down from generation to generation of the hybrids long after the scientists have left.

Natural disasters occur and destroy the stories and artifacts. Technology is gone and the hybrids are left picking up the pieces of their once-great civilization. They find temples and structures from the past and attempt to rebuild them to their former glory but without the help of advanced technology from their god-scientists the structures crumble over time. Where once great monuments stood, the hybrids place primitive stones, attempting to restore them to their former glory. More generations pass and the knowledge of the Genesis scientists has turned from personal knowledge to myths and legends. This could be humanity's story in the distant future.

More importantly, it is humanity's story in the distant past. Thousands of years ago, many different species of extraterrestrial beings came to Earth to seed the planet. Each of these beings created a specific type of human by mixing their DNA with the DNA of their predecessor. Each race had a different genetic marker, causing skin color to be the marker of who the people belonged to and were created by. Each group of humans lived in different parts of the planet and all had contact with these extraterrestrials who lived among them and taught them language, mathematics, sciences, agriculture, and many other things that gave man the ability to move from the caveman era to create a society around him. This is the cultural memory of the Great Leap Forward[a]. These beings presided over each group of men and women, promising to keep them safe and provide for them until it was time to leave and let man rule over himself and others.

Our ancestors wrote stories, created artifacts, and built temples and statues of these extraterrestrial friends to remember this contact and pass the information down through the generations. Most of the information has been wiped from the mainstream narratives of religion. Yet this ancient story of extraterrestrial contact, wisdom, advanced knowledge, and advanced technology has withstood the test of time.

---

[a] The Great Leap Forward is a term used to describe the sudden movement from primitive caveman to civilized man characterized by the rise of culture, art, agriculture, and advanced knowledge from an unknown origin that occured approximately 10,000 years BP.

Mythology: (Greek) Myth—mythos—written or spoken words; *legends and tales of historic accounts sworn to be accurate by kings and priests.*
Logy- *logia; to speak, tell.*[b]

Hidden knowledge buried underground for centuries has risen to the surface and we are beginning to see the connections between humanity and our extraterrestrial creators and friends. From Mesoamerica to the remote tribes of Africa and all across the globe, stories of extraterrestrial contact, advanced technology, and knowledge live in the cultural memory commonly referred to today as myths.

Myths become legends and legends become fairy tales told to children as fun stories. Children are told that there is no truth in fairy tales. Yet, it is precisely in these myths that truth appears. Whether metaphorical or factual, myths may hold the truths we desperately need today. The truth is that we are not alone in this universe and we are not a species of chance.

From the outside, the Black and Native American communities of modern times look like they are two groups of people who were persecuted over the past few hundred years. Today, both of these groups, to a large extent, are dedicated to the Christian message and attend church regularly. They have some ties to their ancient heritage, but it has become whitewashed by Christianity. Where they once spoke of their ancient friends from the stars, they now speak of Jesus and his ability to provide and heal, despite struggling to demonstrate these miracles. What happened over the past two millennia that changed these people's history from oral traditions of contact knowledge to something entirely different?

Through the pages to come, let us take a journey from the valley of the Tigris and Euphrates Rivers to Egypt, Ancient Greece, Rome, Africa, Europe, and the Americas to unveil the true history and unleash the knowledge that extraterrestrial contact and advanced technology have

---

[b] Etymonline.com/mythology

been present since humans first appeared on this planet. The only way to tell this long tale is in multiple parts. Part One is about ancient history and combines various stories to give a broader look at what happened in the distant past. We will look at evidence of ancient technology recorded throughout the world in various texts and archeological discoveries. Part Two tells the modern story that we are told to show the large-scale manipulation over the past two millennia of mankind's ancient past with extraterrestrials. We will also discover that the Christian Church is at the heart of the manipulation and intentional change of ancient history. Part Three tells the grand story of history while uncovering the secrets to see how everything connects to the time mankind is in now and reveals the grand conspiracy the puppet masters have worked towards for the past 4000 years.

It is time to tell a new story of our origins to create a different future for everyone, where secrecy is a foreign word and open discussion is commonplace. Myth belongs to our current reality and has long been forgotten. Let us begin to unpack the creation myths and find the truth of humanity's origin story. Lastly, let us start to understand that rebranding extraterrestrial contact into a worldwide religion has been a carefully constructed ploy of control all along.

Part One
Ancient History

A brief overview of how a worldwide civilization was destroyed, the remnants left behind, and where our modern history begins today.

[illegible] One

An[illegible]t History

A brief over[illegible] [illegible] [illegible] [illegible]lization [illegible] destroyed [illegible]
[illegible] left behind, and where the [illegible] [illegible] [illegible] today.

Ancient Antiquity

One of the seven most incredible wonders of the world is the Pyramid of Giza in Egypt. It stands at 481 feet and is awe-inspiring to every person who sees it. This is only one of the remnants of a worldwide civilization lost to the sands of time. These massive stones used in ancient construction work are called megaliths. They can be seen on every continent, whether in pyramids, temples, or the unexplained existence of megaliths that seem out of place. What if all these megaliths are echoes of a civilization that inhabited the entire planet before a great catastrophe that wiped out most of the population? All across the globe are anachronisms in the form of incredible structures that look older than modern history states they should be.

Most of these megaliths are said to fit within the standard timeframe of 4000 years. Without understanding in-depth geology and carbon-14 dating of sediment and rocks, along with cognitive dissonance, it is easy to understand how these megaliths are so quickly glossed over. Researchers, archeologists, and geologists dated all these monoliths to approximately 10,000 Before Common Era (BCE). This means it is probable that there was a worldwide civilization, that the flood story in all cultures is accurate, and secrets are lost to time. To find these secrets, one must look deeper into these stories and look at megaliths, myths, and legends as a gateway to our distant origins.

In *Timaeus and Critias,* Plato introduces the reader to the story of Atlantis. Critias tells Socrates about a story that was passed down from his great-grandfather Dropides, a friend and relative of Solon. Solon traveled to Egypt and visited the Temple of Sais, where he learned about Atlantis and the once-great civilization from a priest. He heard stories about the high civilization that occupied concentric circular islands and had a high level of technology, which ultimately was destroyed and sank into the Atlantic Ocean. This tale informs the reader that the gods lived among and ruled over man. A war took place, and the fabled Atlantis fell into the ocean. The time of these events was 9000 years prior to the story; these stories were told in 260 BCE so we can place Atlantis' destruction to roughly 11,000 years ago.

Until recently, the story of Atlantis has been a fantastical tale that excites the imagination and generates movies and books. The discoveries within the past hundred years have turned this incredible tale into a probable reality. Atlantis is thought to have been just a three-ringed island. Yet, it was a small part of a civilization covering the globe.

The Mayan flood myth tells of Hurucán or U K'ux Kaj, the creator god who caused the great flood. Native Hawaiian tribes have a figure similar to Noah called Nu'u. Ancient Sumerian stories tell about Gilgamesh and his travels after the great flood in the famous *Epic of Gilgamesh* and the Akkadian epic *Atra-Hasis* speaks of the Anunnaki god Enlil who caused the flood. Egypt relates its flood myth to the gods Ra and Sekhmet. In Vedic tradition, the flood is written about in *Shatapatha Brahmana* and tells of an impending flood that will destroy the world. Europe has many flood myths varying from region to region. Australia holds the myths of the Lizards and Platypuses, China has the Great Flood myth, and African flood myths are told through oral traditions such as those of the Kwaya, Mbuti, Massai, Mandin, and Yoruba tribes. Each tribe within the Americas has a unique flood myth. Lastly, the most well-known flood myth is the story of Noah from Genesis. In all, there are forty flood myths around the world.

This worldwide agreement of a catastrophic flood that changed the earth should be considered more than a myth. The only question is, if there was a flood, is there evidence of something that was wiped away before the flood?

Megalithic Evidence of Civilization Before the Flood

Water is a heck of a force that destroys everything in its wake. If there were to be a catastrophic flood today, how would we see the remnants of today's society? In his book *The Adam and Eve Story,*[1] Chan Thomas writes about a coming cataclysm that mirrors the stories of humanity's ancestors. A pole shift brought this about. In his book *The Earth's Shifting Crust,* Charles Hapgood first theorized the pole shift, and his theory of Earth's crustal displacement was later proven.[2] When a pole shift happens, the earth's spin stops, yet the water and air continue to spin at the 1000mph speed at which the earth rotates. This causes global earthquakes and decimation from 1000mph winds; the molten sublayer breaks through the crust and spews out fire and molten rock, causing all traces of civilization to be destroyed in less than a day. In the movie *The Day After Tomorrow,* a glimpse of this catastrophe is seen. The materials used today to build skyscrapers and monuments would not hold up to this singular catastrophe. There would be nothing left of our society.

Along with the story of Atlantis comes the story of the continent of Mu. This continent existed in the Pacific Ocean and was a landmass larger than Africa is today. It covered the Pacific Ocean from the Japanese Islands to Hawaii, Polynesia, and south-eastward to Easter Island. People of this lost civilization spread throughout the Americas, Hawaii, Easter Island, and the Polynesian Islands. James Bramwell and William Scott-Elliot hypothesize that this continent was destroyed approximately 11,500 years ago, around the time of the sinking of Atlantis. Remember this time frame because it will continue repeatedly throughout this chapter. Traces of this long-lost civilization are seen in the Yonaguni Monument off the coast of Japan, at Nan Madol in Micronesia, and at the Moai megaliths of Easter Island.

Easter Island

Figure 1: Moai Monoliths

More evidence is needed to paint a more credible picture of a worldwide civilization than today's society. Thankfully, there is a plethora of evidence to pull from. At Easter Island, known as Rapa Nui to the native peoples, are the famous Moai Monoliths. These Moai are commonly considered the heads on Easter Island overlooking the sea. However, in 2017, researchers started digging and found the bottom of the statues an astounding 30 feet below the ground. They found that each of the headed statues had complete bodies etched with petroglyphs on the backs of the statues.[a] They were perfectly preserved in dirt and created the perfect opportunity for carbon dating of the organic sediment surrounding the base of the stones. While mainstream archeology places their age at 800 years ago, information from around the globe can lead to a different age range of these monoliths.

---

[a] A petroglyph is a carving or incised drawing on rock (American Heritage Dictionary 5th edition).

Each Moai statue has arms carved at the sides with hands wrapped towards the naval [Figure 1]. This same type of hand placement can be seen on the other side of the world at a famous site called Göbekli Tepe and her sister sites, Nevalı Çori and Karahan Tepe [Figures 2 & 3]. Researchers found statues with the exact hand placement as those of the Moai statues. This hand placement can also be seen on monoliths in Indonesia, South America, Polynesia, and the Korean Peninsula. This coincidence means only one thing: large-scale communication between cultures. The commonly accepted age of Göbekli Tepe is roughly 11,000 years old.

Figure 2: Urfa Man

Credit: Dosseman CC BY-SA

Figure 3: Nevali Cori Statues
Credit: Dosseman, CC BY-SA 2.0, Wiki Commons

The foundation of these Moai monoliths is a platform called *ahu*. While there is a variation of bricks from bottom to top (the same variation seen worldwide), the term ahu is specifically used to identify the megalithic bricks at the bottom of the platform. The word 'ahu' is similar to the Egyptian word *aha*, which is a shortening of the phrase Acu Shemsu

Hor (Ahu Shem-su Hor), which means the Shining Ones and refers to the followers of Horus. The ahu is also found in Machu Picchu, Sacsayhuamán in Cuzco, at the base of the Menkaure Pyramid at Giza, the Osirian Abydos Temple in Egypt, and the stones in Baalbek, Lebanon at the Temple of Jupiter. These connections can be made because of the common characteristics of the stones and the names associated with the platform.

Within the homes of the Rapa Nui people are tablets with an unknown language called Rongorongo. While Rongorongo is a language isolate,[b] similarities exist between it and the script of the ancient peoples of Mohenjo Daro in the Indus Valley of Pakistan. Not until Catherine Ulissey noted that the Rongorongo symbols resembled Anthony Peratt's work with plasma forms did an understanding of the script come to light. This means that the petroglyphs of Rongorongo are possibly an ancient record of plasma formation seen in a type of solar flare activity last seen in 10,000 BCE.[3]

North America

In the early 1900s, G. E. Kincaid accidentally stumbled upon an ancient city within the Grand Canyon while searching for gold, copper, and minerals. He found ancient Egyptian hieroglyphics on the walls inside the cave. Kincaid, Professor S. A. Jordan, and a team from the Smithsonian Institute discovered a complex that could have housed up to 50,000 people. This complex was full of various rooms, including granaries, metalwork, temples, living quarters, kitchens, and catacombs filled with mummies, similar to those in Egypt. Within the center of this complex lies a central room housing a large sculpture resembling Buddha with artifacts that resembled the ancients of Tibet. Today, the entrances to this area are completely blocked off and have become a forbidden zone. However, the rock formations around this zone have interesting names, such as Isis Temple, the Tower of Set, Horus, Cheops

---

[b] A language isolate is a language with no known linguistic affiliation with any other language.

Pyramid, Manu Temple, The Buddha Temple, The Krishna Temple, and the Sheba Temple.[4]

Figure 4: Petroglyphs at Chaco Canyon

In northwest New Mexico lies Chaco Canyon within the Navajo Nation. Known as the Chaco Complex, it comprises 200 settlements covering an area larger than England. Before the 19th century, the Great Houses were the tallest buildings in the United States. There are twelve Great Houses: the most famous of these is called Pueblo Bonito. This D-shaped complex covers three acres, houses 800 rooms, and is five stories tall. Chaco Canyon is stated to have been in use between 900 and 1150 CE. Yet, the petroglyphs found at this site are similar to those found worldwide [Figure 4]. These petroglyphs mirror the ones that Catherine Ulissey noted from the work of Anthony Peratt with plasma formation. The overall construction of stone placement and architecture mirrors Mohenjo-Daro and Göbekli Tepe. While the modern time frame of this settlement is 900 CE, circumstantial evidence similar to the Moai monoliths of Easter Island would place this settlement at approximately 11,000 BCE, yet no definitive date has been made.

Located within the trees of Peebles, Ohio, sits Serpent Mound. It is 1,348 feet long, and an aerial view shows a serpent winding through the trees. The culture that produced this aerial marvel is unknown. Mississippi has the Mississippi Mound Builders; West Virginia has the Grave Creed Mound. Platform Mound is in Georgia, and Cahokia Mound Complex is in Cahokia, Illinois. The flat mound at the Cahokia Complex is ten stories high and covers fourteen acres, with a base larger than that of the Great Pyramid of Giza.[5]

Central and South America

Figure 5: Basalt Olmec Head

Credit Gary Todd, CC BY-SA 2.0,

Mesoamerica holds the memories of the Aztec, Toltec, Olmec, and Mayan Empires. Megalithic Olmec heads were found in 1862 by Jośe María Melgay y Serrano [Figure 5]. These heads show negroid features and are made of basalt block. Since the discovery in 1862, eleven heads have been discovered: four in La Venta, five in San Lorenzo, and two in Veracruz.[6] Each head weighs almost forty tons and is between 6 and 9 feet tall. The Mayan Empire had many pyramids, including the one at

Chichén Itzá and Palenque. Teotihuacan is an ancient city full of pyramids. Within the complex lies the Pyramid of the Sun, the Pyramid of the Moon, and many smaller pyramid structures. The Pyramid of the Sun is 233.5ft tall, with a base perimeter covering approximately 9.75 acres. It is the largest pyramid in Mesoamerica and the third-largest pyramid in the world. The Pyramid of the Moon lies beside the Pyramid of the Sun and is approximately 4.72 acres. Monolithic statues are found in Tula, Mexico, atop a pyramid. These statues are said to be of Toltec warriors, each standing 13 feet tall. While the dating of all these marvels places them within the past 2500 years, the type of construction, placement, use of stones, and sculpting all show echoes heard around the globe.

Machu Picchu sits 7,970 feet high in the Andes Mountains of Peru. Modern scholars state that this site belongs to the Incan Empire and was built in 1400 CE. This city contains over 200 buildings including houses, temples, pathways, and altars. This complex contains stone blocks weighing over fifty tons each and is made of granite containing quartzite.[7] Quartzite is one of the hardest materials in nature, and the precise placement of the stones matches the ahu in Easter Island, the ahu base of the Menkaure Pyramid, and the stones at Sacsayhuamán; these stones fit together so precisely that one could not fit a razorblade between them. The precise dating of this site is unknown, but when compared to cultures around the world, it fits into the 11,000-year-old timeline.

High up in the Andes Mountains of Bolivia at an altitude of 12,800 feet are Tiwanaku and Pumapunku. Megalithic stones in Pumapunku have become known as the H pillars. The largest of the H pillars is 7.81m long, 5.17m wide, and 1.07m thick, with an approximate weight of 131 metric tons.[8] The precision of these stones matches that of other megalithic stones worldwide in Turkey.

Europe

At the north end of the Euphrates River stands Göbekli Tepe. It was accidentally discovered in 1963 and later excavated in 1994 by Klaus Schmidt, who dedicated his life to the excavation of this complex. Large

T-shaped pillars stand up to 5.5 meters tall, weighing up to twenty tons each. There are at least twenty-five small sites that make up the larger complex. While the measurement of these pillars is astounding, the entire site spans a distance of twenty acres, and through excavation, Klaus Schmidt noted that these sites were intentionally buried.

Zooming out from Göbekli Tepe, Karahan Tepe is roughly forty miles southwest, and while Göbekli Tepe is a temple complex, Karahan Tepe is a structured society. Nevalı Çori, Sayburç, and ten other settlements were found in a 100-mile ring around Göbekli Tepe, leading researchers to discover that this region was once an ancient metropolis. Today, this ancient metropolis is known as Taş Tepeler and is 7,854 square miles. Compared to this ancient metropolis, it is slightly smaller than the country of Israel, which is 8,019 square miles.[9]

Northwest of Tas Tepeler lies the area of Cappadocia, where the underground city of Derinkuyu is located. It was accidentally discovered in 1963 when a man was renovating his basement and stumbled upon a chamber. This chamber led to the ancient city and after excavation it was found to be a multilevel complex of over 200 underground cities up to 300 feet below ground. The dating of this complex is approximately 12,000 years old.[10]

In Baalbek, Lebanon, stands the Stone of the Pregnant Woman. This megalith is 20.76m long, four meters wide, and 4.32m high, weighing 1,500 tons. It was originally thought to be the largest megalith in the world until the discovery at Gornaya Shoria in Siberia, Russia. Along a ridge stands a great wall of granite blocks that are three to four thousand tons each, stacked 130 feet high. These blocks have perfectly placed doorways and walkways into the ridge.[11]

India

Using sonar in the Gulf of Khambhat in Western India, scientists discovered outlines of a complex that stretches five miles long and two miles wide. Scientists investigated this region but it is still not recognized in mainstream archaeology due to controversies about the truth of this discovery.[12] This city is fabled to be the mythical lost city of Krishna from the *Mahabharata* and is said to be over 9,500 years old.[13]

Egypt

Egypt holds the most well-known megalithic structures in the world, including the Great Pyramid of Giza, the Sphinx, and the pyramid complex at Giza. Many lesser-known pyramids in Egypt are just as incredible. While some pyramids still hold their original form without the limestone covering, other pyramids, such as the Pyramid of Djedefre at Abu Rawash, have crumbled with time. Djedefre pyramid sits atop a mountain and there is a polished slab of granite and megalithic stones similar to the ahu megalithic stones found at this site. The Great Pyramid of Giza covers thirteen acres and is constructed of 2.3 million stones that weigh up to 200 tons each. This is just one of the three pyramids at Giza that sit in what is known as the pyramid complex. This complex also is home to the Sphinx.

At Abu Gorab stands the base of the largest obelisk in Egypt. It is made entirely of quartz crystal. The base is a large hexagon with perfect precision to the cardinal points. Inside the hexagon base stands a circular disk that is 6 feet in diameter. South of Abu Gorab stands the pyramid ruins at Abusir. Where once stood seven or more pyramids, now stand only three. The famous bent pyramid is found at Dahshur, and the base of this pyramid still has its limestone covering.

The last of the incredible pyramids of Egypt stands the Step Pyramid at Saqqara. Modern Egyptology states that the pyramids were built by slaves around 2500 BCE. Compared to other megalithic sites, it is easy to see that this date and technique of craftsmanship are inappropriate and that the pyramids are much older than modern Egyptology says.[14]

The continent of Africa contains more pyramids, mysterious megaliths, and monoliths than most people know. While the pyramids at Giza are one of the most well-known megalithic structures in the world, they are a tiny fraction found within the continent. Egypt has 138 pyramids from Abu Rawash to The Bent Pyramid in Dahshur. Egypt is just the north end of a vast continent.

Africa

Sudan is home to over 200 pyramids. Known as the Nubian Pyramids and the Pyramids of Meroë, these pyramids may be smaller than Egypt's large megalithic pyramids, but they are still incredible. Modern-day archeologists claim that these pyramids, like the ones in Egypt, were built to hold the remains of the Nubian kings even though no burials or human remains have been found. These pyramids are said to have belonged to the Kingdom of Kush, the land that is today known as Sudan.

In South Africa, Adam's Calendar is a mysterious site known to the locals as The Birthplace of the Sun or *Inzalo y'Langa*. Said to be older than Stonehenge, Adam's Calendar contains multiple megalithic stones and stands atop a mountain ridge overlooking remnants of a forgotten city. The astronomical precision of this ancient site rivals that of Stonehenge, and Adam's Calendar contains megalithic stones that are structured to be an ancient astronomical calendar.[15] The type of stone used to create this ancient megalithic marvel is dolerite, equivalent to volcanic basalt. Volcanic basalt is the same type of stone used for the Olmec heads of Mesoamerica. The original size of these megaliths is unknown due to the extreme weathering that has eroded them over time, and still stand over 7 feet tall.

Adam's Calendar overlooks a valley of remnants that holds the secrets of a lost civilization. This lost civilization is called Great Zimbabwe and is considered the physical location of the Aspû of *Atra-Hasis* and *Enūma Eliš*. These three lost cities in South Africa cover 10,000 square kilometers.[16] The design of the forgotten ruins shows that ancient architects understood Sacred Geometry. Stone walls are 1.5m wide and create shapes such as the star tetrahedron, a wagon wheel or a chromosome, and the phi-factor. The fact that these designs are found in a lost civilization of South Africa shows the world that this lost culture had advanced knowledge of geometry, energy, quantum physics, and astronomy. Michael Tellinger calculated the number of stones that would have been used to create just one of the three lost cities. To create one city covering 10,000 square kilometers, he determined it would need 32.58 billion stones with a total weight of 651.6 billion kilograms.[17]

Just north of Adam's Calendar and Great Zimbabwe lie two pyramids. They are collectively known as Adam's Pyramids and are visible from both sites in South Africa. In his book *Temples of the African Gods,* Michael Tellinger notes that the pyramids and Adam's Calendar lie on the 31 E longitudinal line. This is interesting because the Great Pyramid of Giza lies at 31.13E. Great Zimbabwe lies at 30.93E, and Adam's Calendar lies at 30.91E.[18]

Petroglyphs are also found on large stones once covered by glacial slabs. Some petroglyphs include the Egyptian hieroglyphic ankh, a five-pointed star, a serpent and horseshoe, and a Sumerian cross.[19] The serpent and horseshoe petroglyphs are identical to the symbols found on ancient Greek coins. There are some South African scripts that match those found in the Indus Valley at Mohenjo Daro. When looking at this incredible lost civilization, these stones have no radiocarbon dating or timeframe. This can only be dated by comparing the structures and petroglyphs to others; even this could be inaccurate. The placement of stones, use of similar petroglyphs, and type of roads created in these lost cities are all similar to the construction in Taş Tepeler, Turkey. Could it be that this lost civilization inhabited South Africa during the time of Atlantis, Mu, and the forgotten civilizations worldwide?

Michael Tellinger suspects that the finds in South Africa are much older than the civilizations in Egypt, Sumer, and the Indus Valley. Whether this lost civilization is a precursor to these civilizations or not, its existence and all the similarities found help tell the story of a worldwide civilization. Remnants of these ancient civilizations are also found in the lost inhabitants, whose stories and remains are found worldwide.

Forgotten Inhabitants

"There were giants in the earth in those days; and also after that, when the sons of God came in unto the daughters of men, and they bare children to them, the same became mighty men which were of old, men of renown" -Genesis 6:4 KJV

Archeologists have discovered two anomalies over the past century: elongated skulls and giants. There are various stories of giants within the Bible that will be spoken about in chapter thirteen. This chapter, however, will focus on the scientific discoveries of giant skeletons and myths from other cultures. Native American tribes have many legends of encounters with giants. The Zuni and Pueblo tribes have stories of giants that lived and emerged from the Grand Canyon. The Choctaw in the Mississippi Valley tell tales of a race of giants who exited the Nanih Waiya cave mound in modern-day Mississippi. These giants were said to have white skin and red or blond hair.

The Navajo and Paiute tribes of the Nevada desert tell the same tales of the Choctaw. Sarah Winnemucca Hopkins, daughter of Chief Winnemucca, remembers the story of the red-haired white cannibals 10 feet tall. She remembers the stories about when the Paiute tribe came together with other tribes to defeat the giants. The tribes cornered the giants into a cave and set the cave on fire. This cave is now known as the Lovelock Cave. In 1924, Lovelock Cave was excavated, and a sandal fifteen inches long was found. This would be a size twenty-eight shoe. There were human remains alongside the mummified remains of two giants. One giant was that of a female 6'5" tall, and the other was of a male over 8 feet tall. In 1872, someone found three skeletons, each 8 feet tall, at Great Serpent Mound in Peebles, Ohio. At Spiro Mounds in Oklahoma, a skeleton of a man over 7 feet tall was discovered and was still in full armor.[1]

While there are stories from Native American tribes of giants, skeletons of giants have been found worldwide. The stories of white-skinned, red-haired giants continued from North America to Peru. In Sardinia, a skeleton was discovered to be over 9 feet tall. While fantastical tales of giants and skeletons have been found, many of these stories from the 1900s have been debunked. Yet some stories and legends live in ancestral memory of not only Native American tribes but also of every continent and culture around the globe, including Christianity.

Elongated skulls have been the topic of many famous movies and are incredibly intriguing. Like everything mysterious, they offer possible information into the past. In 1902, an underground temple in Malta was discovered by accident. This temple has since been named the Hypogeum of Ħal Saflieni. During excavation, thirty-three chambers were discovered within three underground layers of the temple. Along with the chambers, over 11,000 skeletons were discovered, of which over 1,000 skeletons had elongated skulls.[2]

Twenty years later, halfway around the world, elongated skulls were discovered in the Pisco province of Peru. Julio Tello discovered the elongated skulls in an elaborate burial site in 1928 [Figure 6]. The burial site was located on a desert peninsula which allowed for the preservation of over 300 elongated skulls. The dating of these skulls is over 300,000 years old. These skulls are the largest in the world and due to their preservation conditions extensive visual and DNA analysis has been performed to show incredible insight into the past.[3]

Figure 6: Paracas Skulls

In 2011, a discovery at the Houtaomuga site in China, would begin the discussion of the practice of intentional skull binding. There were 25 skeletons unearthed at an ancient burial site. These skeletons were carbon-dated and were between 5000 and 12,000 years old. Eleven of the 25 skeletons had signs of intentional skull binding.[4] The age of this discovery links antediluvian culture and religious practices with post-diluvian culture. The practice of deliberate skull binding has been found worldwide in the Americas, Europe, Russia, Australia, and China and is still performed in remote tribes in Africa.

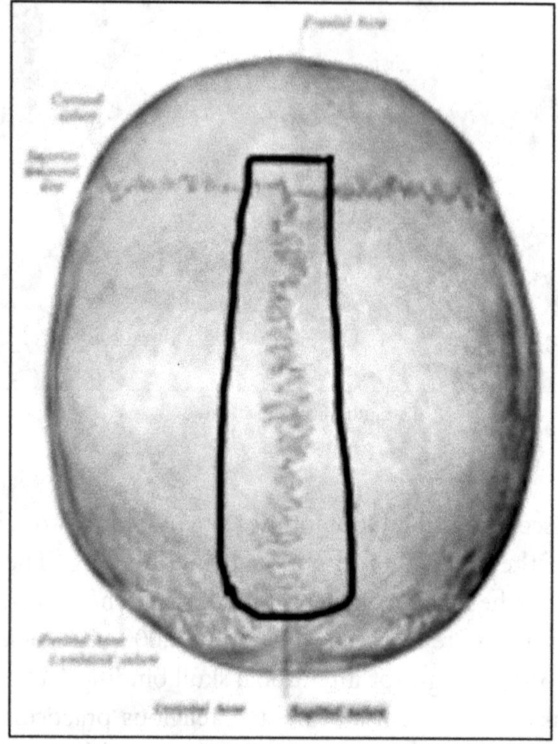

Figure 7: Sagittal Suture

Credit: Public Domain, Wiki Commons

Intentional skull binding and elongated skulls have been grouped, yet the two have pronounced differences. A human skull has two parietal plates with a sagittal suture forming the center of the skull [Figure 7]. The Paracas skulls have only one parietal plate and an absent sagittal suture.[5] An ongoing study analyzed the results of the John Hopkins Genome deep sequencing project in 2019 and traced the elongated skulls found in Paracas, Peru, to the Black Sea region along Malta.[a] The most surprising finding was that the mitochondrial DNA on the moth-

[a] Genome Comparative Analysis of Skulls from Paracas Region of Peru with 1000 Human Consensus Reference (Genome: Nassim Haramein & William Brown, August 2020).

er's side was not closely related to *homo sapiens sapiens*, the anthropological classification of modern humans, and the Y chromosome of the father's DNA shows no correlation to *homo sapiens sapiens.*[6] The elongation of the skull allows for a larger cranial capacity of up to two and a half times that of the skull of a modern human.

Petroglyphs Found Across the World

The Nazca Lines in Nazca, Peru, cover hundreds of kilometers and extend over 500 square kilometers, or 190 square miles. On the ground level, these lines look like natural indentations of the earth. They are meant to be seen from an aerial view, and looking down, they resemble giant versions of petroglyphs. These Nazca lines are attributed to the Paracas culture. Petroglyphs found at this site include a monkey, a condor-like bird, a hummingbird, a pelican, a spider, a whale, and various plants, trees, geometric shapes, and spirals. These massive petroglyphs are similar to petroglyphs found on stones worldwide.

In Ireland, there is a place called Loughcrew tombs, locally known as *Sliabh na Caillí* (Mountain of the *Cailleach*). It consists of three hills with peaks reaching over 900 feet high and covering over 100 miles. Only ten percent of sites have been discovered. Each individual site is made of stone and called a cairn. There are two main cairns (Cairn T and Cairn L), with over thirty other sites still standing. It is theorized that there were between 50 and 100 structures when this unknown civilization existed.

Figure 8: Carin T Petroglyphs

Credit: Voxlivia CC: BY-SA 2.0, Wiki Commons

The cairns are thought to be burial grounds, places of celebration and gathering, and temple-like complexes, with each cairn serving a different purpose. The largest cairn, Cairn T, is 38 yards across (456 feet), and a large carved stone called Hag's Chair with thirty-seven other stones sits along the base of the cairn. Within Cairn T are stones covered in petroglyphs like the one in Figure 8. Each individual cairn contains petroglyph-covered stones, and archaeologists think that the stones were once covered with white quartz.[7] A fascinating non-coincidence is there is an equinox stone in Great Zimbabwe that is covered with petroglyphs mirroring these ancient ruins. The state of the stones at the Loughcrew site also reflects the natural weathering seen at Great Zimbabwe.

What does this all mean then? One needs to put all the pieces together to form a cohesive picture. There was a worldwide civilization that shared culture. This civilization traveled the seas and built megalithic monuments and statues. They were a civilization from the stars and seeded this planet. These beings with elongated skulls lived among giants and humans. A great flood came and wiped away this grand civilization. The cause of the great flood is unknown, but it changed the face of the planet. Remnants are left of this story and live in the ancestral memories of tribes and cultures worldwide.

Advanced Knowledge: Star Alignments of Megaliths

There are 40 flood myths spanning 4000 years over six continents. The probability that the flood is a myth is understandable when only one flood myth exists. Two flood myths decrease this possibility, but forty myths signal a stop, pause, and a deeper look into what history is saying. Why are there so many flood myths? Why are megalithic sites popping up all over the world with dates older than the flood myths? Why would ancient civilizations go through so much trouble to build all the incredible megalithic structures in chapter one, only for them to be wiped out by a flood?

The megaliths of the antediluvian civilization tell stories that society is only now beginning to unravel. Each site has precise astrological alignment pointing the way to the stars. It is as if they were created to remember the home of those who came and helped humanity in times of trouble. Whether it is the sun, planets, or other stars out in the universe, a connection to the stars is at the foundation of all societies. Ancient Egyptians worshipped the sun with their god, Amun. In *That Old Time Religion*, Jordan Maxwell, with co-authors Paul Tice and Alan Snow, writes about how religion formed from the beginning of society. A stellar, lunar, and solar-based belief system is still seen today. Astrology is an ancient practice based on the zodiac that dates back to

Babylon and ancient Greece.[1] Yet these regions' stories, practices, and teachings come from places much older, Ancient Egypt, whose culture is carried over from Atlantis.

The cosmos held such an essential role in antediluvian civilization that almost every megalithic site on the earth has a connection to the stars. In *The Orion Mystery*, Robert Bauval details how he discovered that the pyramid complex at Giza was an exact representation of the belt stars of Orion.[2] Each of the three large pyramids is perfectly aligned with the three stars that make up Orion's belt and this perfect alignment is seen during the vernal equinox. This discovery gave way for Wayne Herschel to find astronomical alignments not only in Egypt but all over the world. In his book *Hidden Records I*, Herschel intricately details the correlation of pyramids and structures worldwide to three major star systems. Others joined in with their research and came to the same conclusion of three major star systems and megaliths. These three major star systems are the Pleiades, Sirius, and Orion. The incredible part is, while the megaliths all align with the star systems, ancestral narratives found all over the world also all refer to these star systems. Even the book of Job in the Old Testament refers to all three-star systems in two verses.

"Can you bind the chains of Pleiades

or loosen the belt of Orion?

Do you bring out the constellations in their season

or guide the Bear with her cubs?"

– Job 38:31-32 TLV

The "Bear with her cubs" references the Ursa Major constellation known as the Great Bear. The brightest star within this constellation is the Sirius star system.

The Sirius Star System

Sirius is the brightest star in the night sky and one of the most influential stars in ancient narratives. Since Alvan Graham Clark's 1862 discovery, it is a scientific fact that Sirius is a binary star system, meaning that it has two stars that orbit each other. But the European and American Scientific communities once thought Sirius was a singular star system. It took another century for the possibility of a third star to be seen. There is debate regarding the validity of a third star versus a planetary body, but modern astronomists' discovery of the second star, Sirius B, verified the echoes from ancient mythologies thousands of years ago.

The importance of this star in ancient narratives cannot be overstated. Robert Temple remarks in *The Sirius Mystery* that Sirius is the star on which the Egyptian calendar was based. The placement of Sirius in the night sky with the rotation and rising of the stars created a twelve-hour time system for Egyptians.[3] The Dogon tribe of Mali, West Africa, are modern-day secret holders of Egyptian knowledge.[4] The Dogon priest spends his life in a cave to preserve the ancient teachings from the outside world. Archaeologists discovered the Dogon tribe was aware of the three stars within the Sirius star system in the 1930s.

Herschel details the star map of Egyptian pyramids along the Nile River starting with the pyramid at Abu Rawash and ending at the pyramid at Mazghuna, just as the star map of the cosmos along the Milky Way begins at Sirius and ends at Canis Major.[5] In Abu Rawash, there are two megalithic pyramid ruins found. The larger pyramid that stands on a mountain, the pyramid of Djedefre, correlates with Sirius A. The smaller pyramid located east of the mountain, called the Lepsius I pyramid, correlates with Sirius B. Karl Richard Lepsius was a Prussian Egyptologist and archeologist in the 19th century. He led a Prussian expedition through Egypt to explore and record ancient Egyptian civilization. During this expedition, his team uncovered 67 pyramids; at Abu Rawash, his team documented three ruins. While only two ruins are seen today, the three ruins are reminiscent of the three possible stars of the Sirius system.

In the Maltese Islands, over 1000 elongated skulls were discovered in the megalithic underground temple of Hypogeum of Ħal Saflieni. This is but one of many temple complexes on the Maltese Islands. At the Origins Conference in 2020, Lenie Reedijk gave a discourse about the Tarxien temples of Malta. These temples are named after the gigantic stones they were built from, and the naming is intentional and based on the belief that only giants could have placed the stones in their current formation. Sixty-six megalithic temple sites once stood atop the Maltese Islands. Of the 66 temples, only 19 were preserved, with each temple showing clear orientation points.[6]

When attempting to discover the orientation points of these temples, researchers found none of the temples face the waters of the islands or directly face the sunrise at the equinox or solstice. Each temple has a specific but irregular orientation when compared to the others. George Agius and Frank Ventura discovered an apparent clustering of orientations in temple door alignments. Through statistical analysis, they realized that the possibility of this happening through coincidence was one in a thousand.[7] Building on this discovery, Reedijk looked toward the stars and found that all the temples have a particular alignment facing Sirius' helical rising and setting in the night sky. Reedijk factored in Precession of the Equinox or Axial Precession[a] and realized that Sirius was seen with the naked eye around 9500 BCE, and the first time that it was seen in the temples was around 9200 BCE, or 11,000 years ago. She discovered that while all the temples have a slightly different orientation towards Sirius, it is as if they follow the precession of Sirius as it dances across the night sky from east to west.[8]

THE CONSTELLATION OF ORION

Orion is one of the most recognized constellations in the night sky. Orion's belt includes three main belt stars that are the easiest to spot, even with the ambient light of a busy city. The three belt stars are Alni-

---

a Precession of the Equinox is the movement of the celestial equator with respect to the fixed stars and the path of the sun's motion in space as viewed from earth (Encyclopedia.com).

tak, Alnilam, and Mintaka, the smaller, slightly offset star of the three. Yet the constellation of Orion is made of many more stars, which creates the Orion constellation. This constellation reflects the god Orion holding a club, lion, or shield. The stars within the Orion constellation that form the body of Orion include Saiph, Rigel, Betelgeuse, Bellatrix, and Meissa, with separate stars that make up the club and the lion. While most of the megalithic sites worldwide follow the simplicity of the belt star alignment, other sites incorporate the entire constellation and create intricate maps with alignment to the other two major constellations.

Starting from the Egyptian star map at Abu Rawash and traveling south to Giza, you will find the famous pyramid complex. Robert Bauval discovered and covered in great detail the alignment of the belt stars with the three pyramids in *The Orion Mystery*. The Great Pyramid correlates to Alnitak, Khafre's Pyramid correlates to Alnilam, and Menkaure's Pyramid correlates to Mintaka. Herschel builds on Bauval's discovery and suggests that two ruins outside the main pyramid complex correlate with Betelgeuse and Rigel.[5] He also correlates the placement of the Sphinx at the pyramid complex to be facing towards the constellation Leo.[9] Herschel notes that a trail of faint stars inside the belt stars of Orion mimic a cosmic serpent.[10] The serpent was highly correlated with wisdom and knowledge and is regarded in Native American, Mesoamerican, and Vedic traditions as a deity.

Chapter one notes the astronomical alignment of Adam's Calendar in South America and shows that this structure is an ancient astronomical calendar. Three megalithic stones are next to a bird-shaped Horus statue along the outer ring of the calendar. These three stones along the outer ring of the circle make up the belt stars. Two other stones within Adam's Calendar make up the top half of the Orion constellation (Bellatrix and Betelgeuse).[11]

Traveling to the opposite side of the world in Mexico stands the pyramid complex at Teotihuacan. Herschel notes that the Pyramid of the Moon correlates with Alnitak, the Pyramid of the Sun correlates with Alnilam, and the Feathered Serpent's Temple correlates with Mintaka. This particular alignment is a mirror reflection and not an exact representation of the belt stars of Orion. Herschel theorized this was because there was a larger plan at foot. He notes that the placement of the Temple of the Feathered Serpent faces the rising of the Pleiades.[12] Teotihuacan is said to be the city of the Aztecs. As chapter one suggested, the Aztecs may have stumbled upon a much older civilization and built upon what already existed.

In northeast Arizona, descendants of the Anasazi tribe, modern-day Hopi, organized their villages to match ancient star alignments. Gary David provides intricate details of this placement in his book *The Orion Zone*[13] and the possible causes of the variation of stars correlating with specific villages. David notes that the three mesas—the flat-topped hills upon which their villages sat—align with the three belt stars of Orion. Orion's right and left shoulders correlate with Bellatrix and Betelgeuse. Orion's knee correlates with Canyon de Chelly, and Orion's foot correlates with Betatakin. In Figure 9,[b] the stars that make up the lion are also shown as a shield of six stars.[c] The placement of the shield correlates with the Grand Canyon, where the Anasazi people lived during the time between the Third and Fourth Worlds with the Ant People. The incredible placement of Orion on Earth correlates to the mysterious Chaco Canyon and Sirius.

---

[b] Credit: Northstar, "The Hopi Remnant" (The North Star Chronicle, 11 Mar 2020. www.thenorthstarchronicle.com/2020/03/11/the-hopi-remnant/. Accessed 8 Dec 2023.).

[c] A secondary interpretation of what Orion is holding in his hand.

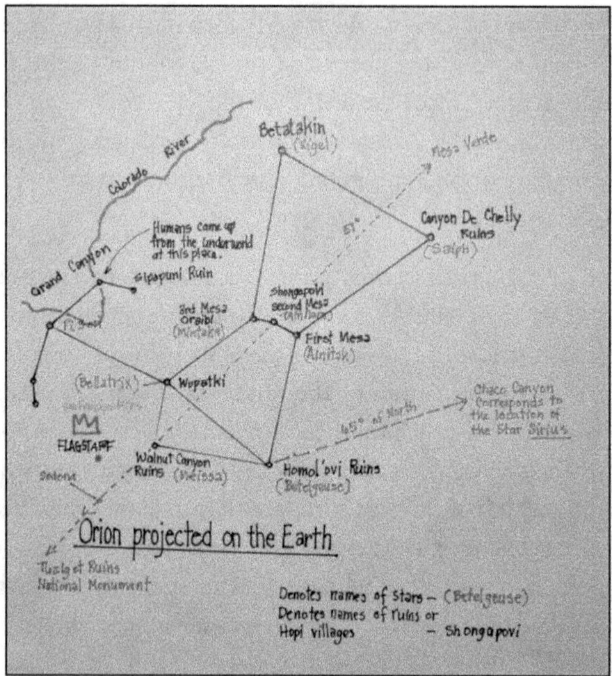

Figure 9: Hopi Mesas and Orion Correlation.
Credit: Northstar

Mauro Biglino is a biblical scholar and translator who supervised the translation of 17 books of the Old Testament for Edizioni San Paolo, Italy's foremost Catholic publisher. In *Gods of the Bible,* Biglino dives deep into the original Hebrew and Greek translations of common biblical terms to reveal their true meanings and origins back to extraterrestrial contact knowledge. He remarks on the Orion and Pleiadean star systems. Chapter fourteen will dive deeper into the biblical commentary about the giants and nephilim. Yet, Mauro points out that while the Greeks translated the word "nephilim" to "giants," the Aramaic use of the word Nephila, the singular form of Nephilim, is a proper name of the constellation Orion.[d] He continues that the knowledge of their

---

[d] *Brown Driver Briggs Hebrew and English Lexicon* gives the Strong's concordance number 5303: giants and notes the Aramaic origin is Orion.

origins is seen in the knowledge of Sirius and its correspondence to Saiph, the star opposite of Sigel as noted in the Hopi mesas.[14]

THE SEVEN SISTERS OF PLEIADES

The Pleiades star cluster is a group of stars located in the northwest constellation of Taurus. Reference to the Pleiades is found in Mayan, Aztec, Native American, Babylonian, Indonesian, Maui, Hawaiian, South Pacific Islanders, and ancient Egyptian stories. This star system consists of over 800 stars, with six main stars seen today. These six main stars are referred to as the Seven Sisters in mythologies across the world. Why six stars for seven sisters? Astronomers used the Gaia space telescope to answer this age-old question and found that the star Pleione is so close to Atlas that it appears as one star to the naked eye. The team of astronomers used their knowledge of star movement and rewound the clock to find out when the Pleione and Atlas would have been seen as two separate stars to the naked eye. They found that this would have been 100,000 years ago,[15] which means that 100,000 years marks the beginning of the seven sisters' stories.

The sky map of Egypt travels from the pyramid complex at Giza and continues south along the Nile River. The following major pyramid site is Zawyet El Aryan, which correlates to Aldebaran.[5] Traveling south along the Nile River is the Abusir pyramid complex. As noted in chapter one, Abusir holds a complex of pyramid ruins that once had seven pyramids, representing the seven stars of the Pleiades. Egypt is the only place in the world where the entire star map of the Milky Way is represented on land.

Traveling 7,519 miles to the jungles of Guatemala stands the pyramid complex of Tikal. Tikal is thought to be the earliest pyramid site of the Mayan Empire. At Tikal stands the Tikal Temple, along with other megalithic structures. The temple complex is a cluster of megalithic sites and Herschel argues that the temple complex correlates with the Pleiades.[16]

In Peru, the Pleiades were revered by the Inca Empire. Within the Machu Picchu complex stands the Torreón or Temple of the Sun. This temple is built into the bedrock of the Andes Mountains. Two trapezoidal windows face the valley below and were placed to watch the Pleiades in the night sky.[17] The precise alignment of the trapezoidal window is seen during the June (winter) solstice when the sunrise perfectly frames the stone inside the temple. This window also frames the movement of the Pleiades.[18] Ricardo González Corpancho notes in "The Pleiadians" that there is a Peruvian temple called Chavín de Huántar that contains an altar called Choque Chinchay that has circular indentions within the stone. He notes that when it rains, the altar fills with water and is meant to reflect the Pleiades.[19]

While few megalithic alignments correlate with the Pleiades compared to Orion, this constellation and the beings who hail from the Pleiades hold the highest respect among ancient cultures. Chapter four explains that these beings aided many Native American tribes and assisted the Mayan ancestors.

The Greek myth associated with these three stars, as Biglino notes, is Orion, the great hunter who was always with his dog Sirius. He was in love with the Pleiades, yet Artemis had fallen in love with Orion, who had him killed by a scorpion. Zeus finds out what his daughter Artemis has done, strikes the scorpion down, and places Orion in the night sky.[20] This is just one of many mythological stories of the constellations, likely due to the fact that Sirius, Orion, and the Pleiades are visible along the horizon through the night.

Ancestral Narratives

Flood myths and creation myths are found all over the world. Yet these all tell the rehabilitation stories of this planet after it experienced immense catastrophe or chaos. A genuine creation story would tell something different. A set of criteria is needed to establish a real creation story. These criteria would be:

- Begin before time existed
- Creation of the universe
- Age of the universe
- Detail the formation of galaxies, solar systems, black holes, and planets
- Explain dark matter
- Possibility or reality of other life forms outside of Earth
- Understanding of the multiverse

This list of criteria shows that the Bible's creation story is not a true creation story but a rehabilitation story. The first ten chapters provide an incredibly condensed version of the stories seen in the Mayan and Hopi stories of the fifth world. The most detailed creation story is that of the Dogon Tribe of Mali, West Africa. Many creation stories begin with the earth already formed and in chaos.

AFRICA

The Dogon of Mali, West Africa

The Dogon Tribe of Mali, West Africa, was discovered in 1935 by French anthropologists Marcel Griaule and Germaine Dieterlen. After spending 15 years with the tribe, the anthropologists were given sacred information by the blind tribal elder named Ogotemmeli. Ogotemmeli told the anthropologists their oral stories of the gods that came from the stars. The anthropologists were amazed at the type of advanced knowledge of the Dogon, including the atom, string theory, the four quantum forces, subatomic particles, and quantum physics that have only been understood in the last hundred years.[1] They knew Sirius A, Sirius B, and Sirius C. They were astonished by the knowledge of Sirius B, which they call *po tolo,* a tiny seed found in West Africa. They noted that Sirius B is a tiny star but the heaviest thing. Surprisingly, Sirius B is a white dwarf star with one of the highest density of stars known. Sirius C was discovered in 1995.

Ogotemmeli also shared the Dogon's creation story. It begins with Amma, the creator of the universe, and their egg, which symbolizes a primordial body that existed before the creation of the universe; it is also referred to as the original placenta. This egg held the potential seeds of future creation and gave the locations of Orion, Sirius, the Pleiades, and a comet. Amma broke the egg and released a whirlwind that scattered matter in all directions. This matter formed into galaxies, stars, and planets. The Dogon even describes the moon correctly as a dead and dry body.

Dogon mythology is based on the principle of twin births. The process of creation began with the imperfect union of Amma and Earth. This union resulted in the jackal's birth, symbolizing disorder and Amma's difficulties. Water is the divine seed in the Dogon tradition, and with the womb of the earth, the perfect pair of twins, the Nommo, was created. The Nommo are named O Nommo and Ogo. They are shaped like fish but have arms and legs and live in water. The Nommo comes from Sirius, the land of the fish, which travels on the Milky Way and lands on Earth on a day called "the day of the fish." The Nommo, as a group, came from Sirius to Earth to give knowledge to man.[2]

Nigeria

Yoruba Creation Myth

Life only existed in the sky, and there was a supreme being called Olorun/Olodunmare. They had a lot of deities that worshiped him, and they had everything they needed. One of the deities, Obatala, wanted to create and looked down from the sky and saw a vast ocean. He received help from the other deities, made a long golden chain, and gathered all the sand in the sky into a snail shell. He added baobab powder, took the sacred egg containing the personalities of the seven deities in the sky, and traveled down the sky for seven days. When he arrived at the ocean-filled Earth, he poured the sand, creating it.

Edo

The great Osanobua [God] asked his four sons in heaven to choose whatever gift of nature they wanted. The oldest chose wealth, the second chose wisdom, the third chose mysticism, and the fourth was about to make his choice until he was interrupted by the Toucan, Owonwon, who cried out and told him to settle for a snail shell. They went to Earth and found water everywhere. They were in a canoe and drifted with the power of the wind until they came across a tree in the middle of the water.

The youngest got out of the canoe, turned the snail shell upside down, and brought the sand up from the bottom of the water. The four elements of creation were made: water, air, fire, and land. The youngest Osanobua was given a physical human body because he stepped out of the canoe. The other sons remained deities, but the youngest was the ruler of the earth. He represents innocence and is susceptible to the powers of his brothers. The oldest brother chose to take his gift and became the god of the water. The two middle brothers disposed of their spiritual selves and gifts and stepped onto the land.

The youngest brother renamed himself Idu [the first human on earth], explored the land, and asked his father for food and a human companion. Osanobua told him to take sand from the ground with both palms and stretch his hands together in front of him. Osanobua called forth a female, pointing his staff where she appeared.

NATIVE AMERICAN TRIBES

Hopi Indians

The Hopi Tribe, located in northern Arizona, is just a tribe of Native Americans. Yet to those who have spent time studying the Hopi and their ancestorial narrative, it is a tribe of immense information, all leading back to origins with the "Star People," as the Hopi call them.

There was endless space, with one being named Taiowa. Taiowa then created Sótykang and then commanded them to create life. Sótykang created nine universal kingdoms: water, air, and water. He then made Spider Woman on earth, blessing her with knowledge and wisdom. She started twins for balance on both poles of the earth. Spider Woman then gathered four colors: yellow, red, white, and black, mixed them with saliva, created their form, and covered them with a white substance cape. These beings took on their own form and were called men. She then made four female partners for the four male forms. Man was created in the dark of purple light, the first phase of the dawn of creation. The second phase was yellow light, and consciousness entered man. The third phase was a red light when man was fully formed and faced his creator.

The man was unable to speak, and thus, language and speech were given to man by Sótykang; each man's color was granted a different language. The people of the First World lived in harmony with nature and each other and maintained a connection with the divine. They understood the chakras and lived harmoniously with their body and the earth.

Over time, the people forgot their connection to the divine and became more focused on the differences between their speech and skin color. Division happened, and war began between the various skin colors. A select few within each skin color group were still in tune with nature and the divine and spoke with Sótykang. When he came, it sounded like a mighty wind. He told them that Taiowa had seen the state of affairs of this world and was going to destroy it. They were going to send a cloud to follow by day and a star to follow by night and to follow them until they stopped. The people were led to a large mound where the Ant People appeared. The people were informed to learn from the Ant People about agriculture and living harmoniously amongst one another. The Ant People gave the chosen people refuge. The first world was destroyed by fire.

Once the destruction was over, Sótykang appeared, leading the people out of the anthill into the second world. Mankind had everything they needed, lived in small communities, and traded with each other. Yet they wanted more. Quarrels and war broke out again. Sótykang again appeared to a chosen few, leading them to the Ant People. The second world was destroyed by ice.

Humanity then emerged into the third world. Man had multiplied, and countries and civilizations were created where villages once stood. Technology had advanced, and man was making pátuwvotas, objects created to fly on. Men used these to attack each other in war. Sótykang and Spider Woman spoke with one another. They agreed to save some and destroy the rest with water. The people Sótykang and Spider Woman desired to save were placed in a sealed hollow reed and given food. The great flood came, and the people waited until the right time to emerge and begin again.[3]

The Hopi knew and called their friends the *Anu* (Ant) *Naki* (Friends). Ricardo González Corpancho remarks that the Hopi speak about the gods of the Pleiades. These gods saved their ancestors from a great catastrophe in the Pacific Ocean. They came in flying shields and took them to different parts of the Americas[4]. Robert Morning Star is a part of the Hopi Tribe. He was raised on a story told to him by his father. His grandfather and some friends rescued an extraterrestrial being whose ship was destroyed. This being was naked Bek'ti and told them of a great story.

The Terra Papers[5,a]

There was nothingness in the void of space. Primordial essence exploded, and, like the Big Bang theory, galaxies emerged after a long period of time. One galaxy in particular was called Eridanus. Within the galaxy Eridanus, planets formed, as did life of all kinds. Morning Star remarks that there were fish humanoids, bird humanoids, insect humanoids, reptile humanoids, and mammal humanoids. Humanoid was the consistent effect of evolution. Primordial humanoids developed into civilized humanoids with advanced technology. They created a space-faring society and civilization among the stars. War began, and what could have been civilized battles became the science of war. Kingdoms rose and fell for millions of years until one empire rose to power with their mind-reprogramming science of war. This kingdom was known as the ARI-AN reptilian star race (pronounced Orion) and were descendants of their dinosaur ancestors. This was a female-ruled race, with the Orion queen ruling ruthlessly. The Orion Empire reigned for thousands of years as a new race of wolf humanoids called KANUS developed from Sirius. Orion and Sirius joined in an alliance: Orion controlled everything, while Sirius' beings were the most feared and fiercest warriors that existed. Together, they controlled the superstar highway within the Eridanus Galaxy, linking the inner and outer star systems.

A new system emerged at the edge of the galaxy. One of the Sirian kings claimed it for himself. Greed and power were the game, and war continued as Genesis scientists made this star system habitable. A perfect planet was created to be a place for the king to rule within the system and mirror the Sirius empire, Tiamat. A coup occurred between the Sirian kings, split between two star systems, and a second-in-

[a] Author's Note: care has been made to tell this story briefly and without copyright infringement. This entire story is fascinating and worth the read.

command named ZuZu overthrew King Anu. By attempting to overthrow Anu, ZuZu went against the Orion, Sirius Empire, and alliance. Both empires were out for total retaliation and destroyed precious Tiamat to destroy ZuZu. Anu, his Genesis scientist son Ea, and Enlil returned to Sirius.

Millions of years later, it was found that from the ashes of Tiamat's destruction, a new planet had emerged. Anu decided control over this planet would be shared between Ea and Enlil and dispatched Ea to gather a group of volunteers and give them titles of lord with land on this new planet. This group was named Annunaki (meaning givers of life) and, with the help of Ea, would make this unique planet habitable again. When Ea and the Annunaki arrived, they had nothing but trouble because the planet's energy grid was destroyed. They built energy centers with four reflective triangular sides to balance the planet's energy. This new planet would be named Eridu in remembrance of what it was.

Once the planet was stable, Ea and Ninhursag worked together to create hybrid creatures for the planet. One hybrid creature was created to be a half Sirian and half Eridu beast. The being was named Adapa and was created to be a slave race to Enlil, who ruled over the sky. Enlil was the secondborn of Anu, unlike Ea, the firstborn. Enlil was born of Anu and his half-sister, and under Sirian rules of succession, Enlil was heir to the kingdom of Sirius and thus ruler of Eridu. Enlil made Adapa slaves through mind control, and his labor force was created to do the work of the Annunaki. Ea went behind Enlil's back because he cared for his creatures and detested how Enlil treated them. Ea believed that a controlled life was not an evolved life. He gave Adapa knowledge of free will and procreation, and when Enlil discovered this, he was furious. He separated the evil transgressors from the loyal slaves, gave them new names, and kicked them out of the garden called Garden for Life. The transgressors were named Adamus, and the loyal subjects, Adapa, were given four commandments: (1) total obedience, (2) all traces of Ea were removed, (3) no sounds or evil sounds against Enlil, (4) obedience lessons every seventh day.

Ninhursag and Ea worked together to find groups of Adamus beings, giving them advanced knowledge of agriculture and the universe's secrets. Enlil learned of this and became enraged. He was going to use his planet-sized ship to destroy Adamus; he told Ea that no one was to know about this upcoming catastrophe. Enlil would take his Adapa into a ship and wait out the catastrophe above the earth while the Adamus were destroyed. Ea defied his brother and told Adamus. He sent a group deep into underground caverns with the Hen-T beings, another group was sent to the mountain highlands, and others were placed in a special cargo ship to wait the storm out.

The Orion and Sirius empires were told of the massive destruction of Eridu and were enraged, forcing Enlil and Ea to a truce. Peace reigned until Marduk, son of Ea, allied with the Orion rebels who lived deep underground on Eridu. They joined together and formed a coup, taking control over Eridu. Enlil fled to Sirius, and Ea fled with many followers to his star system, Pleiades. Marduk destroyed all traces of monuments created and all tablets that told the tale before Marduk ruled. He reprogrammed all Adamus and Adapa forcibly. Knowing their fate, the followers of Ea still on Eridu fled to the wilderness or mountains and engraved Ea's teachings on tablets of stone for a future time. After all beings were reprogrammed using mind control war science, Marduk gained total control, and the Sun God Ra was born. Ra created a doctrine to support Ra, but the teachings of Ea became evil, and ancient history was now a myth or fable to the newly reprogrammed beings.

One particular note of interest that Morning Star makes in his story is that the Sirius beings called the Annunaki (including Anu, Ea, Enlil, and Ninhursag) all had large cranial lobes, nostrils, and ear flares, similar to the elongated skulls found across the world.

Other Native American Tribes

The Hopi are not alone in this belief that Star Beings created them. Native American tribes all over North America, including the Cherokee, Lakota, Onondaga, and Cree, all believe that they came to earth by beings from the cosmos.[6] The Cherokee of East Tennessee believed their creators were the Pleiadeans, beings of light. These beings rode a sound wave through a hole in space. This hole landed in East Tennessee, and the Cherokee Indians were the guardians of this vortex.[7] Native American stories are filled with tales from the stars. The story "The Origin of the Pleiades and the Pine"[8] tells the pine story and references the pinecone shown throughout the world in ancient artifacts. They also speak of these beings giving advanced technology to man.

Mesoamerica

Popol Vuh

While this book is not mainly about Mesoamerican culture, the information it contains helps to understand its relation to other cultures worldwide, especially when attempting to grasp the idea of Earth rehabilitation. The *Popol Vuh* is the creation story of the Mayan culture and holds valuable information from Mesoamerican cultures that have been lost to time.

The story begins with everything being silent and calm when only water and sky existed. The Heart of the Sky (Huracan) arrived with the Quetzal Serpent and Sovereign. They took away the water and then created land and mountains. After the gods created the animals, they made man. They used mud, but there was no substance for man, and they had no mind of their own, so they destroyed them with a flood. The second attempt at the creation of man was with wood. These creations were immortal, did not honor the gods, and were destroyed in hot water. The Farmer and Shaper attempted to create man out of maize, and they had a grand vision and could see and understand everything. The gods realized the creation had become like gods but could not replicate themselves. The gods decided that the essence of man would be reframed and reshaped. Women were created and men's vision was blurred so they could not be like the gods.

All the gods gathered to await the arrival of the first dawn when the sun, moon, and stars would be seen for the first time. Before the sun came up for the first time, the face of the earth was wet and soggy. The face of the earth, the sky, and dawn became clear. A few humans rejoiced with the gods as the world dried.

The creation stories within this chapter give a more in-depth explanation of the Genesis creation story. Could it be that the first ten chapters of the Bible provide the summarized version of these stories to avert attention away from other cultures while simultaneously still telling them?

The Bible and The Sages

The Bible

While the following chapters will focus extensively on the Bible, an alternative view of the first two verses of Genesis is necessary. The Christian tradition holds that all creation was made in six days. What if the creation story in Genesis One is a story of rehabilitation, and the original translators chose specific words to use during their translation and skewed the truth of the words used?

Genesis 1:1

"In the Beginning created God the heavens and the earth"

"bə-rê-šîṯ bā-rā 'ĕ-lō-hîm; 'êṯ haš-šā-ma-yim 'êṯ hā-'ā-reṣ".

The words to be retranslated in this verse are:

- Beginning – [בְּרֵאשִׁית] *bə-rê-šîṯ*, meaning first phase, step in course of events.

- God – *ʾĕ-lō-hîm* (*Elohim*), meaning rulers, judges, and divine representatives.
- Created – [בָּרָא] *bā-rā*, meaning to shape, create, form, or fashion by cutting.
- Heavens – [הַשָּׁמַיִם] *haš-šā-ma-yim*, meaning sky or skies.
- Earth – [הָאָרֶץ׃] *hā-ʾā-reṣ*, meaning land.

Retranslated: "In the first phase of events, the Elohim shaped the sky and land."

Figure 10: Recreation of the Sumerian explanatory pictogram seen in Biglino. *Gods of the Bible*, p. 68

Genesis 1:2 speaks about the "Spirit of Elohim." The Hebrew word for spirit is [רוּחַ] *ruach*. Biglino captivates this visual representation of "spirit" or *ruach* perfectly. The Hebrew origins are found in the Sumerian language, and the sound [*ru-a*] has a pictogram, as seen in

Figure 10. This pictogram contains an object (RU) above water (A). He recalls that even though visually this pictogram makes sense, to understand its deeper meaning, he borrows the acronym RIV from the *Lexicon Recentis Latinitatis* and remarks that *Res Inexplicata Volans* [RIV] exactly translates to "Unexplained Flying Object."[a] The other word to consider is [מְרַחֶפֶת] *merachefet* (*mə-ra-ḥe-p̄eṯ*). This word is only partially translated. The rest of the word is the participle of the verb. Biglino notes the verb "hovering" and the participle "to shake or vibrate." [1]

Genesis 1:2

"And the earth was formless and void and darkness [was] over the face
wə hā-'ā-reṣ, hā-yə-ṯāh ṯō-hū wā-ḇō-hū, wə-ḥō-šeḵ 'al- pə-nê

of the deep and the spirit of God was hovering over the face of the waters"

ṯə-hō-wm; wə rū-aḥ.[ruach]. 'ĕ-lō-hîm, mə-ra-ḥe-p̄eṯ 'al- pə-nê ham-mā-yim"

The words to be retranslated in this verse are

- Was – [הָיְתָה] *hā-yə-ṯāh*, meaning come to pass
- Formless – [תֹהוּ] *ṯō-hū*, meaning formless, confusion, reduced to primeval chaos
- Void – [וָבֹהוּ] *ḇō-hū*, meaning of primeval earth.
- Face – [פְּנֵי] *pə-nê*, meaning edge.

---

[a] *Lexicon Recentis Latinitatis: Volumen 1.* Et 2. Urbe Vaticana, Libraria Editoria Vaticana, 2003.

- Spirit – *rū-aḥ* [*ruach*], meaning wind. RIV: Unidentified Flying Object.
- Hovering – *mə-ra-ḥe-p̄eṯ*, meaning to hover and shake or vibrate.

Retranslated: "And it came to pass, the land was reduced to primeval chaos. There was darkness over the edge of the water. And the UFO of the Elohim was hovering and shaking, suspended over the water's edge."

The first book of Genesis records what was done to rehabilitate the earth after a planet-changing catastrophe. Verse three speaks of light. Verses four and five references separating the light [day] from the dark [night]. Verse six records the separation of fresh water from salt water, with a surface between. Verses seven and eight recall separating the water of the sea from the water of the sky. This verse reflects the sentiment found in the *Popol Vuh* that the earth was soggy. In verses nine and ten, the waters of the deep were gathered together and called a sea, while the dry land became visible and was called the earth. Verses eleven to thirteen speak of vegetation. Verses fourteen to nineteen record the sun, moon, and stars, similar to the *Popol Vuh* and the witnessing of the first dawn.

Using the criteria for a creation story from chapter four, it is easy to see that the story in Genesis 1 is not a creation story at all but one of rehabilitation. Rather it begins with the formation of man and woman, millions of years after the formation of planets.[2] In fact, Genesis 1:1-2:4 is a compilation of the Babylonian *Priestly Code* from the 6th century BCE.[3]

The Seven Sages

When reading ancient creation stories, one group consistently mysteriously arrives to Earth, gives humans advanced knowledge, and then leaves. This group is known as the Seven Sages, and they travel to different groups worldwide and serve one master or lord. In Mesopotamia, the Seven Sages are called Apkallu, meaning wise, sage, or expert. There are seven demigods whom Enki created, and before the flood, they served the kings and priests. In the *Epic of Erra*, also called *Erra and Ishum*, Enlil speaks to Erra and asks where the Seven Sages of the deeps are. He refers to the Seven Sages as sacred fish with perfect wisdom like their lord, Ea.[4]

Figure 10: Oannes Credit: Dr. Regosistvan CC BY-SA 4.0, Wiki Commons

The *Bīt mēseri*[5] is an incantation text on four cuneiform tablets. It describes the Seven Sages and says they come in three forms: ūmu, fish, and bird Apkallus. The Seven Sages originated in the river and, after the flood, were banished to the Apsû. This text gives the names of the Seven Sages:

- "Uanna: who accomplished the plans for heaven and earth.[b]
- Uannedugga: who gives understanding.
- Enmedugga: whose good fate is decreed.

[b] U-Anna (Sumerian); Oannes (Babylonian)

- Enmegalamma: who was born in a house.
- Enmebulugga: who grew up in a pasture.
- Anenlida: the conjuror from Eridu.
- Utuabzu: who ascended to heaven."[6]

Hess and Reiner agree with the correlation between Utuabzu in the *Bīt mēseri* and Enoch of the Bible.[7] Berossus, in *The Chaldean Account of Genesis*, comments on the Seven Sages and states that the first of these sages is named Oannes [Figure 11], who comes from the water to give knowledge.[8]

Figure 11: Above: Apkallu holding flowers in their hand with watches around both wrists.

Credit: Shukit Muhammed Amin CC BY-SA 4.0, Wiki Commons

Left: Apkallu holding the handbag

In reliefs, the Sages are shown as half men, half fish, or a man with a fish covering. They have a handbag in their left hand, a pinecone in their right hand, a watch on both wrists, and sometimes a bracelet, mace, or flowers in their hands, as Figure 11 shows. The handbag represents the tools needed to create civilization, the pinecone represents knowledge, and when pointed in one direction, it represents the passing of knowledge and wisdom.[9] The depictions of Sages who hold the pinecone and handbag symbolize the ones who gave this ancient knowledge to man, and the passing of this information could be the foundation for the

Great Leap Forward. There are many references to Asakku, yet when used in the context of the Mesopotamian stories, they refer to figures like the Apkallu and are usually references to gods when they have been defeated.[10] Asakku is also referred to in the *Atra-Hasis* as the disease that was sent by Enlil but is specifically paired with the name of the disease (*šuruppu*)[c].

In Sumerian Epithets of Enki, he is depicted as Ushumgal, a serpent with great wisdom. The Vedic texts speak of the Nāgas, who are divine serpent beings and are half human, half serpent. They inhabit the subterranean water kingdom. In Mesoamerica, their leader is depicted as a feathered serpent holding a handbag with their left hand. Along with the megalithic Olmec heads found in La Venta was a stele of a man holding a bag in his hand. In Egypt, the leader is Osiris, Lord of the Underworld, whose followers are called the "Followers of Horus" or "The Shining Ones."

In Dogon's stories, these sages are the Nommo who live in a water reservoir and are teachers and masters of water. In the Incan tradition, there are seven helpers called Hihiwapanti. In *Ancient Civilizations,* Freddy Silva remarks that Hihiwapanti is from the Aymaran language and translates as *The Shining Ones.*[11] O'Brien comments that the '*El'* in Elohim has the meaning of 'interest'. Comparing ancient world translations, *El* translates to *The Shining One,* and Elohim translates to *The Shining Ones.*[12] Enoch recounts the verbal description of a few of these Shining Ones when he notes the beings who came into his room and woke him up: "Their faces shone like the sun."[13]

This direct correlation is made of Yahweh.

c The *Atra-Hasis* notes: Enlil told Namtar to send the šuruppu, ašakku, blow into them like a storm (Dalley, *Myths of Mesopotamia*, p. 23). Yet, Sitchin clarifies that *NAM* is fate that cannot be altered, while *NAM.TAR* is destiny that can be altered (Sitchin. *The Cosmic Code*, p. 61, 88-89).

"Restore us, O Yahweh: let your face shine, that we may be saved!"

– Psalm 80:7 ESV

"I lifted my eyes and looked, and behold, a man dressed in linen with a belt of fine gold from Uphaz around his waist. His body was like yellow jasper, his eyes like a flash of lightning, his eyes like fiery torches, his arms and his feet like the gleam of burnished bronze." – Daniel 10:5-6 ESV

Whether the Shining Ones were in Mesoamerica, India, or Egypt, or written by biblical editors, this all cooperates and shows evidence of a worldwide civilization that existed and communicated with one another long before modern-day humans thought it was possible. It also shows there were beings whose skin shone and who were not from this world.

The connection between the antediluvian civilizations of Atlantis and Mu can be traced through one character common to both: Thoth. Thoth is depicted as a being with the head of an ibis bird and the body of a man in ancient Egypt. The *Bīt mēseri* notes that an unknown number of the Apkallu is found in bird form. The *Emerald Tablets of Thoth the Atlantean* states this connection is evident when the tablets are opened.

"I, THOTH, the Atlantean, master of mysteries, keeper of records, mighty king, magician, living from generation to generation, being about to pass into the Halls of Amenti, set down for the guidance of those that are to come after, these records of the mighty wisdom of Great Atlantis."[14]

Thoth explicitly states that he brought the wisdom of Atlantis to the people of the new age after the flood. He and the other Apkallu brought wisdom from before the great catastrophe to the age after. The next chapter will detail how Thoth appears in various texts across cultures, specifically the Bible, and how his true story was erased from the Biblical narrative. This knowledge has again come into the light of day with the resurgence of Coptic texts and Dead Sea Scrolls.

Part I Summary

New archeological evidence discovered over the past hundred years has caused man to look at history anew. What was once considered a myth has become a fact due to the overwhelming evidence challenging modern archaeology regarding the existence of an ancient civilization that predates the great deluge. When myths found worldwide correlate with one another, the very idea of myth must leave and be replaced with hard evidence. Megaliths and petroglyphs provide proof of a new story. Extraterrestrial beings came to Earth, created man, and lived among humanity.

As Part II will show, this narrative was written in cuneiform texts for future generations to discover and unravel the truth. A flood occurred, and forty different traditions told stories about the flood. Within each story is a correlation to other stories found across the world, particularly the stories coming out of Mesopotamia. The Anunnaki helped man to survive because they created mankind. Researchers Arun Durvasula and Sriram Sankararaman studied the ancestry of West Africans. They found across all Neanderthal, Denisovan, and modern humans that there was a "ghost" sequence of DNA with unknown origin. They also found that West Africans must have come from a population that placed their DNA into Neanderthals and modern humans.[1] This is scientific proof of ancient ancestral memory echoing throughout time.

The names of humanity's extraterrestrial ancestors changed throughout the regions and ages, but their memory remains in modern myths, epics, and mankind's DNA. The first ten chapters of Genesis tell the condensed version of the creation epics of Mesopotamia, also known as the Generations of Isis (*Gen-i-sis*). Another story emerges that blends fiction and history, causing a confusing and inaccurate representation of what really happened after the great deluge.

The gods became known to man while one reigned supreme. Cultures shifted, names changed, and languages were transformed throughout time, causing the identity of the extraterrestrial ancestors to be lost. A once great story of antediluvian knowledge and megaliths was changed to create a cohesive story that no longer makes sense.

Part Two

Extraterrestrials, Advanced Technology, Giants, UFOs, Egyptian Pharaohs, and Twisted Narratives

How ancient ancestral narratives and accurately written extraterrestrial contact narratives were twisted and contorted to create a monotheistic society and stamp out memories of our otherworldly ancestors and helpers.

Changing the Biblical Narrative

Ancestral narratives or myths often present as existing in a world far removed from our own. Modern-day American society often looks at textbooks and the Bible to create a cohesive history. The modern-day Western system of years is based on the birth of Christ and is treated as the timetable upon which civilization is based. This creates a timeline of approximately 6,000 years within the established allowable historical record of civilization. Adhering to this strict timeline within the mainstream conversation does not allow for new discoveries to be made that lie outside of the 6,000-year time limit. When attempting to understand ancient history, it becomes necessary to look outside the mainstream information readily given and seek sources from societies and megaliths long forgotten and recently recovered.

The oldest civilization known to man is Sumer. Between the Tigris and Euphrates Rivers, Sumer holds the oldest records of civilization. These records were discovered at the beginning of the 20th century by the German American scholar Hermann Hilprecht when he found the first fragment of the *Sumerian King List* cuneiform tablet. This discovery opened the way for all future researchers to seek an alternative history of mankind. The first line reads, "After the kingship descended from heaven, the kingship was in Eridu."[1] The tablet continues to tell the story of five cities and the eight kings that ruled that region for 241,200 years before "the flood swept over." After the flood, kingship was again lowered from heaven.

After initially reading this, a spark of curiosity was born that turned into a wildfire because of the phrase "kingship descended from heaven." The wording can only mean one thing: someone else came from the stars, down to earth, to rule over man. Since that first discovery by Hilprecht, hundreds of tablets have been found to work together to create a very different history than man has been led to believe. The story of *Atra-Hasis* was rediscovered in the Library of Ashurbanipal. It was written in cuneiform on clay tablets and was later translated by George Smith and Stephanie Dalley. The *Atra-Hasis* and *The Epic of Gilgamesh* are said to be the oldest stories known to man. What is interesting about the *Atra-Hasis*[2] is that it is a creation story like the one found in Genesis but with much greater detail.

UNDERSTANDING THE BACKGROUND OF THE BIBLE

What would you say if you found out the Bible talks about extraterrestrial contact? The response would probably be somewhere along the line of "Roderick, you are crazy."

For centuries, Christians have been led to believe that the Bible is the infallible word of God, incapable of being wrong. No matter the denomination of the Christian Church, it is taught that the Bible is the literal word of God that man wrote down. The general term used is "divinely inspired," as if the authors of the Old Testament sat at a desk as a ghost-like form of God stood over them and dictated to them a history lesson word for word. God's ghost would then hover over translators and tell them how to translate history from one language to another or something along those lines.

Yet the Bible is just a compilation of stories that were orally transmitted for centuries and eventually written down sometime around the 4th century BCE. The creation stories of the Sumerians, Babylonians, Egyptians, Olmecs, Dogon, Mayans, and many other cultures were also written down long after their creation. It is taught that the lineage of Abraham wrote most of the Old Testament, yet the author of the first five books—Genesis, Exodus, Leviticus, Numbers, and Deuteronomy—is unknown. No matter who wrote the text, it was a human being who was no different from anyone else. The original writers wrote down stories to remember their past, the same as we do today in textbooks. Just like what appears in modern-day history, the story of the past is written by the victors, not the victims. The stories of ancient societies and belief systems that have been destroyed have been lost to time. To find the truth of mankind's past, the "divine or inspired word of God" label needs to come off. It needs to be reread through new eyes as if man was experiencing things they could not understand and simply chose to write down their observations. These observations can be used today to create a new framework and make new conclusions about previously held beliefs.

One factor that makes it difficult to understand that the Bible is a compilation of stories about extraterrestrial contact is the various translation tactics that have been used throughout the millennia. Integral words that describe advanced technology and extraterrestrial contact have been serendipitously transformed to create a monotheistic system that controls the followers through fear and manipulation.

Instead of believing everything in existence was created in six literal days, reread Genesis chapters one and two as if the writers were using readily available information. This information would have been the stories from cuneiform tablets. The mysterious writers of the Old Testament would have been well-versed in previous creation stories. The Tigris and Euphrates rivers run through the fertile crescent; within the fertile crescent lay Assyria, Nineveh, Babylon, Sumer, and Eridu. West of the fertile crescent is Mount Sinai, where Moses met with God face-to-face and received the Ten Commandments. These stories were told for centuries before the invention of the Hebrew language around the 10th century BCE. The writers grew up hearing the stories of otherworldly beings living among man, and a long, lengthy explanation or creation story was not needed. Consider

Genesis chapters one and two as the summarized version of Sumerian or Babylonian creation stories.

Cuneiform and Ancient Texts

Instead of just reading the summarized version of the creation stories the Church has spoon-fed, follow the trail of the 10,000-year timeline given by megalithic structures and begin in Sumer. The stories were orally transmitted for hundreds if not thousands of years and then written down in cuneiform on clay tablets. If history had been written on paper, it would have been destroyed with time. Sumerians wrote in cuneiform on clay tablets to prevent this destruction and preserve their history. As with *The Code of Hammurabi*, Sumerians also recorded bills of sale, daily activities, and their type of civilization and government. They also told the stories of their ancestors' lives with extraterrestrials. It was recorded knowledge and not just a fictional story as it is thought of today.

The art of cuneiform is tedious, and a series of wedges are pressed in clay to create characters. Cuneiform uses 600-1000 characters and creates words by diving up syllables.[a] The writers of ancient extraterrestrial contact painstakingly took the time to intricately detail their history so that it could be passed on. Just like the Rongorongo script of Easter Island, cuneiform was a language islet, and it was not until the studious work of George Smith that it could be successfully translated into the modern-day language. With the help of Sir Henry Rawlinson and Sir Austen Layard, Smith could dedicate his life to this work and reveal the secrets of our past.

Understanding the Myths

In the ancient stories, the names of the different gods vary throughout regions, societies, and cultures. While the names are important and distinctive, they are not as important as the roles and characteristics

---

[a] BritishMuseum.com

given to the character. The change in names is commonplace throughout the ancient world. Beginning with the Sumerian epic *Atra-Hasis*, An is the father god who rules over the sky, Enlil is lord over the earth, and Enki is lord over the waters (underworld, deep, or Aspû). To completely understand this story, it is broken up into small pieces with an explanation.

The Creation of Man

Atra-Hasis begins with the Anunnaki gods, who are hard at work digging canals and separating waters to make the land inhabitable. The Anunnaki gods are unhappy because they are overworked and decide to make the lower gods, Igigi, do the work instead. The Igigi dug the Euphrates and Tigris rivers, raised mountains, and created agriculture. After 3,600 years, the Igigi revolted against Enlil because the work was too hard, and attacked his home. Enlil appealed to his father, An, for advice on what to do; Enlil was told to send Enki to find a solution. Enki arrived at their side and devised a plan to create man to do the work. Enki worked with his sister, Ninmah, to create man through cloning. The *Atra-Hasis* shows that fourteen clay pieces were mixed with a sacrificed god's blood. The fourteen clay pieces were split; seven were placed on the right and seven on the left. Ninmah then cut open the umbilical cords. After ten months, man was born.

Zechariah Sitchin analyzed this Sumerian creation story in *The Twelfth Planet*[3] and noted that the clay is specifically mixed with the blood of a god. The first man and woman were created using gene manipulation, surgical procedures, and in vitro egg fertilization. Enki and Ninmah used the already existing female *Homo erectus* genes to mix with the genes of the male gods (Anunnaki).[4] After the creation of *Homo sapiens sapiens* was successful, they were cloned in batches for mass production. Enki got tired of all the work so he tinkered with genetic manipulation and enabled man and woman to reproduce.[5]

. . .

Yes! That right! An extraterrestrial named Ninhursag of the Anunnaki gave birth to man!

That means that *Homo sapiens sapiens* look like the Anunnaki. Zechariah Sitchin recounts Marduk's words in *The Lost Book of Enki* when Marduk falls in love with a human. He asks for permission from his mother, Ninki, and father, Enki, to marry a human named Sarpanit. He recounts that humans are made in their own image and their own likeness. This means that humans have everything except the longevity of the Anunnaki.[6]

The first set of *Homo sapiens sapiens* were clones, and if one clone died, they must be replaced with another. After getting tired of the clones dying off, Enki and Ninhursag returned to the drawing board, did some tinkering, and adjusted the clones to reproduce independently. Genetic markers were then placed on their creation so that they could be distinguished from other groups of people.

This is seen in Genesis chapter four after Cain kills Abel, Cain is spoken to by God (or Yahweh) and is cast out from the area where he lived. Cain states:

"Whoever finds me will kill me." Yahweh rebukes him and says, "'Not so; anyone who kills Cain will suffer vengeance.' [Yahweh] then put a mark on Cain so that no one who found him would kill him." – Genesis 4:14-15 ESV (adjusted for understanding)[7]

Billy Carson notes that the mark on Cain is a DNA marker seen on the skin, known as skin color or race, meaning that each race resembles the ones that created the race and ruled over them.[8] All throughout the stories of the Anunnaki, mankind is referred to as "black-headed people." In the *Code of Hammurabi*, Hammurabi says that he was given the right to rule over the black-headed people like Shamash.[9]

Six hundred years after a man was created, Enlil became enraged because man was too loud. He then ordered disease to decrease the number of men on Earth and to quiet them down. Enki's son Ziusudra lived on the earth. He pleased Enki as a way to appeal to Enlil and save the people. Ziusudra was instructed to build a temple for Namtar,[b] the secondary god of the underworld, and the disease was wiped away.

Another six hundred years passed and man was too loud again. Enlil was once again angry and sent a drought to stop the growth of vegetation and kill man by starvation. Ziusudra went to Enki again and was told to build a temple for Adad, son of An, to stop the drought. This appeal did not work, and mankind plundered into cannibalism. The Anunnaki were still furious with man and made a pact to send a flood. Enlil made it known among the Anunnaki that no man was to learn of this flood, and no *Homo sapiens sapiens* was to survive. Enki went against Enlil and told his son Ziusudra to build a boat to save his bloodline and son. The flood came and killed everyone. The Anunnaki wept, and Ninhursag held the Anunnaki accountable for their actions. Enlil spotted Ziusudra's boat, became irate and called for An. An discovered that Enki was the one who saved Ziusudra and stated that he preserved life in defiance of the Anunnaki. Mankind was cursed, and women were to have difficulty in childbirth.

To understand the names, gods, and deities associated with the Anunnaki and their reference in the Bible, the variation of names throughout the region of Mesopotamia is needed. Traveling northwest from Sumer to Babylon, the names begin to change slightly. An, Enlil, and Enki remain, and the audience is introduced to Bel-Marduk (often called Bel, Ra, and Marduk). Ziusudra's name changes to Utnapishtim. Continuing northwest on the journey is Akkad, where the biblical city of Nineveh is found. The name of An was changed to Anu, Enlil to Ellil, Enki to Ea, and Utnapishtim to Atrahasis. In Akkad, the oldest stories known to man, *The Epic of Gilgamesh* and *Atra-Hasis*, are found.

. . .

---

b Namtar is often referred to as the god of death but Sitchin remarks that it is a type of fate that can be altered and not a god.

Multiples of God?

In the Bible, some of these Anunnaki gods are mentioned. Genesis 1:26 points out that multiple gods were present during the creation of man: "Let us make man in our image, in our likeness." Even the Bible reveals that man looks like the ones who created him. This means, of course, that mankind looks like an extraterrestrial. This verse is often glossed over after hearing it and reading it many times, or the Church explains that *our image* refers to the triune nature of God. The Old Testament was written in Hebrew, and to fully understand this implication, the original language offers the solution to the question of who is "our."

The Hebrew text uses the word Elohim for God in this verse. Elohim is the plural of Eloah. It makes sense when this is put into the context of the Anunnaki and Igigi. However, calling extraterrestrial life forms gods is an inaccurate description. They were not gods; they were bipedal humanoid beings from other planets, and as chapter two revealed, they possibly had elongated heads, and some were giants. Yet their possession and use of advanced technology caused modern man (at the time) to worship them as gods.

Genetic Material or DNA?

The original Hebrew words reveal that man was made in the image of the gods, which gives more information and is strikingly similar to the *Atra-Hasis*. Genesis 1:27 says that the Elohim created man in his own image.

- Image = [צֶלֶם] *tzelem*
    - Strong's Concordance No. 6754 – image, construct. A type of abstraction.
    - *Etymological Dictionary* – something material that contains the image or complete form.[10]

Today, we know that it is DNA that contains the image and exact likeness of a specific human. This biblical narrative mirrors the *Atra-Hasis* and notes that the gods used the purified blood of an Anunnaki male to create man.[11] This biblical narrative is also directly mirrored in the *Lost Book of Enki* when Enlil speaks to his three sons. He states, "In the beginning the Earthlings in our image and after our likeness we made."[12] It is as if the translators of the biblical narrative directly quoted Enlil and placed the Earthlings part within Genesis 2:7.[13] If the Bible openly stated Earthlings, it would bring up questions that the church is still not truly ready to answer.

This biblical reference points out that each group of beings that seeded Earth used different genetic markers. The *Thiaoouba Prophecy* comments that the Bakaratini arrived on the earth 1.35 million years ago with two yellow and black lineages. The yellow lineage arrived in the region between China, India, and Thailand, while the black lineage traveled to Australia and Africa. The beings that traveled to Africa were the separatists, and their stories are found within the stories of the Anunnaki.[14]

It is also crucial that all the Sumerian, Akkadian, and Biblical stories are of the Anunnaki. The original location of the Anunnaki is known today as Great Zimbabwe and shows the ruins of the once-great empire, as noted in chapter one. The DNA from the Elohim or Anunnaki would also have the coding or genetics of skin color, meaning that the Anunnaki had black skin. George Smith gives this argument solid ground. His translation of three fragmented creation stories shows the evidence as plain and simple. Ninmah speaks about the newly formed man "in the mouth of the dark races which his hand has made."[15]

Smith notes that this race is the dark race (*zalmat-ququdi*) and is referred to in other fragments as Admi or Adami. He points out that Sir Henry Rawlinson commented during their discoveries of the ancient cuneiform tablets that the Babylonians acknowledged two principal races: the Adamu, or dark race, and the Sarku, the light race.[16] This means that the Babylonians were aware of both Caucasian and Negroid races. So much so that the references made in Genesis about the creation of man were specifically the creation of the Negroid man. Therefore, this can only mean one thing. If a man (in the region of Great Zimbabwe, also known as Aspu) was made out of the DNA of the Anunnaki or Elohim, then that means that they were Negroid. Thus, the entire Genesis story is about the creation of the Negroid race.

Wait! What?!

Yes, this information is completely subversive to the narrative that exists today.

The Thread of Mankind's Creator

In *The Terra Papers*, Bak'ti informs the listener that when Ea created *homo sapiens sapiens*, his secret intention was to create a race of beings that would be greater than the gods. He took a select group of the Adapa[c] and gave them self-awareness and an understanding of beauty in the world without the influence of oppressive systems designed to control them. He wanted to give this new being named Adamus the ability to belong within the larger universe as well as create their own destiny. Ea also gave the Adamus beings the desire to pursue sexual desires over work. News of this new being made its way to nearby star systems and galaxies. The Akhu[d] sent Ea the gift of passion in a single filament of DNA to add to Adamus. This gift of passion allowed Adamus to have more passion, emotion, and feeling than their overlords would ever understand. This gift was called The Gift of the Feather. Yet both the Adamus and Adapa were given free will.[17]

Adamus and Adapa both had the gift of intelligence, and correlating various stories together, Ea worked with these Adamus beings to create the Brotherhood of the Snake. It was a mystery school open to both Adamus and Adapa. Ea and Thoth taught the scientific approach to spiritual evolution. The spirit was not a thing of mysticism but practi-

---

[c] This term was used by Morningstar as the general term for *homo sapiens sapiens*.

[d] A bipedal bird species known today as avian.

cality that could be understood at a scientific level. The levels were open to everyone, but to move from one level to another, the student had to master the lower level. Each level was kept a secret by an oath of the student to avoid harm to the student. Similar to a student who has a basic level certification as a nurse aide and wants to go to medical school, there are designated steps toward their ultimate goal. The structure and teachings were effective and became known as the Ancient Mystery Schools of Egypt, and before the arrival of the second Pharaoh Khufu (also known as Cheops) in the 4th dynasty, they thrived.[18]

When Khufu became pharaoh, his full name was Khnum-Khufu, after the Ram god Khnum. He built the first temple in 2900 BCE as a house for the first mystery schools. This marked the turning point of the Brotherhood of the Snake. The Kheri Heb (High Priest) was the only one who knew the mysteries and secrets and shared them only by word of mouth. No documentation was ever written down. The secret wisdom was to be transmitted only to those who knew. Architecture and hieroglyphs were the early forms of transmission.[19] The teachings that were given in these mystery schools perverted spiritual truth, focused on the scientific, and ignored the mystical. While the original mystery schools knew that the spiritual was as easily understood as any other type of science, this new school taught the opposite.[20] The knowledge of extraterrestrial ancestors began its dark course through history.

The Beginning of the Biblical Patriarchs

The Adam and Eve story of Genesis is told in the *Atra-Hasis, Myth of Adapa,* and *Enki and Ninmah: The Creation of Humankind.*[1] After Adam and Eve, Cain moved east to Nod, and Seth was born. Six generations later, Enoch is born to Jared. In Genesis, the life of Enoch is given three verses stating that he fathered Methuselah and walked with God; after three hundred years, he had other sons and daughters, and when he was 365 years old, he was no longer because God took him. This is the only story of Enoch that is given—or is it? When the Dead Sea Scrolls were discovered, the *Book of Enoch*[2] was found among many other writings called the *Apocrypha*. Within the Dead Sea Scrolls, some stories are vastly different from the curated texts of the Old and New Testaments. The reflective reader is shown a different version of Jesus and an entirely different version of history. Suppose Church teachings intend to sway Christians from asking too many questions about life outside of Earth. In that case, it is no wonder why the *Book of Enoch* was not included in any modern translation of the Bible other than the Ethiopian Bible.

Unknown Books of Enoch

The *Book of Enoch* tells tales of the Watchers, extraterrestrial abduction, levitation, portals, traveling vast landmasses very quickly, and space travel. The *Book of the Secrets of Enoch* reveals the appearance of the beings who visited him. He notes that they woke him up and took him away.[3] It even references the nephilim that are spoken about in the Bible. '*Nephilim*' is a Hebrew word that translates to 'giants.' Unsurprisingly, there is another apocryphal text called *The Book of Giants*. The Watchers came down from heaven, and they

"lusted after [the daughters of man], and said to one another, 'Come, let us choose wives from among the children of men and beget us children.'" - Enoch 6:2

Along with starting families, the Watchers lived among the children of men. They taught humanity alchemy, jewelry making, meteorology, agriculture, astrology, astronomy, and solar and lunar orbits. The depth of knowledge of even one of these subjects is advanced; to teach humanity all of these things is incredible.

The passages in chapters seven and eight of the *Book of Enoch* plainly state who gave mankind the Great Leap Forward. No wonder this book was left out of the carefully canonized Bible. Daniel 4:13 even remarks on the Watchers and states that one of them "came down from heaven."[4] Mauro Biglino comments that this is a possible reference to the fact that the Sumerians called their country the "land of the Watchers" (*Kiengir*), and the Egyptians knew of the Watchers from ancient times.[5]

The Lost Book of Enki: Ties to Enoch

Various Mesopotamian stories and tablets reveal more information about the Anunnaki gods' lineages. Sitchin remarks that Enki was named Ptah in Egyptian stories, and the group he arrived with was named *Ntr*, meaning 'guardian' or 'watchers,' showing that these stories traveled throughout the entire ancient world.[6] Tablet eight of *The Lost Book of Enki,*[7] the Adam and Eve story, is told from the Sumerian point of view. An incredible statement is made by Enki, who tells Endubsar that during Enshi's life, the Anunnaki became lords over man.[8] The *Bible* perfectly echoes this in Genesis 4:26: "A son was also born to Seth, who he named Enosh. Then the people began to call on the name of Yahweh." These two insights perfectly align and show that Anunnaki or Elohim existed among man but did not rule over them until a specific time. As future chapters will show, the most incredible fact is that this specific Elohim, initially one of the worker gods, took over and decided to rule some two hundred and thirty-five years after Adam or Adapa was created.[9]

The biblical character of Enoch is known as the scribe of judgment or great scribe. In *The Lost Book of Enki* tablet eight, Enki-Me[10] is the seventh in lineage from Adapa, and Sitchin notes that Enki-Me is taught the months by the moon's movement and the years by the sun's movement. He makes two celestial journeys, is taken by Marduk in a rocket ship to the moon, and is taught by Utu the function of priesthood and how to begin the priesthood. Enki-Me recorded all his journeys and knowledge, then gave the stories to his son Matushal.[11] Like Enoch of the Bible, Enki-Me is remembered in the annals of the heavens, where he departed and lived until the end of his life.

Hebrew Name	Sumerian Name in *Lost Book of Enoch*
Adam	Adapa
Eve	Titi
Cain	Ka-in He Who in the Field Food Grows Taught by Ninurta/Nimrod
Abel	Abael He of the Watered Meadows Taught by Marduk
Seth	Sati He Who Life Binds Again
Enosh	Enshi[a] Master of Humanity
Kenan	Kunin Learned Alchemy
Mahalalel	Malalu He Who Plays [music]
Jared	Irid He of Sweet Waters
Enoch	Enki-Me By Enki ME Understanding
Methuselah	Matushal Mighty Man[b]

Table 1: Names of Biblical Characters correlated with Names from The Lost Book of Enki

Enmendurana,[a] also known as *Enmeduranki's Story*, is found in

[a] Enmendurana (also Enmeduranki) 7th king on *The Sumerian King's List*. EN.MEN.-DUR.ANA means "Master of the Divine Tablets of the Heavenly Bond." EN.ME.-DUR.AN.KI means "Master of the Divine Tablets of the Bond Heaven-Earth." *En* means 'chief,' who reigned in Sippar for 21,600 years (Sitchin. *Divine Encounters*, p. 64-68.).

writings by Wilfred Lambert.[12] This story states that Enmeduranki, King of Sippar (formerly Zimbir), was taken to heaven by the gods Shamash[b] and Adad,[c] where he was taught the secrets of heaven and earth, just as Enoch is in the *Book of Enoch*. Enmeduranki is given a divine tablet with the secrets of heaven. These secrets of heaven include information about the solar system, the constellations, knowledge about geography, and geometry, just like Enoch. It was Enmeduranki whom the Anunnaki gods groomed to be the first priest among the people.[13] Enmeduranki can be seen as Enoch's Sumerian counterpart, just as Noah's Sumerian counterpart is Ziusudra.[14]

Other names of Enoch include Ningishzida, the son of Enki, who was sent to Earth with Enki by Anu to assist man with knowledge and build civilization. Enoch is called Thoth in Egypt, and he is called Hermes Trismegistus (meaning 'thrice great') in Greece. Thoth was Egypt's great teacher and knowledge keeper and Hermes was the great teacher of Greece. As seen in chapter five, no matter the name, he helped shape the ancient world.

Enoch is the father to Methuselah, who is the father of Lamech. Lamech is the father of Noah. Noah is the tenth great patriarch of the Genesis story, and Xisuthrus is the tenth great king from Berosusus' *Chaldean Account of Genesis*. Each tenth great ruler or patriarch is associated with the great deluge. In *The Epic of Gilgamesh*, Gilgamesh searches for Utnapishtim to find the key to immortality. Utnapishtim found this key after the great flood (Utnapishtim/Atrahasis is Akkadian while Ziusudra is Babylonian). In *The Lost Book of Enki*, it is revealed that Ziusudra is the son of Enki by Batanash, the wife of Lu-Mach. Enki does not want his offspring to be found, and with the help of Ninmah, he is born and raised in Shuruppak, then marries Emzara and has three sons.[7] Ziusudra's kingship is found in the *Sumerian Deluge Myth,* and he is called the king and gutug-priest.[15]

---

b Shamash is the divine judge. Associated with the underworld, he serves as the god of divination.

c Adad is the weather god.

Sumerian King List (Old Babylonian)[8]	Genesis Genealogy	*Chaldean Account of Genesis*
1. Alulim	1.Adam	1. Alorus
2. Alalgar	2. Seth	2. Alaparus
3. Enmenluana	3. Enosh	3. Amillarus
4. Enmengalana	4. Kenan	4. Ammenon
5. Dumuzi- the shepherd	5. Mahalalel	5. Daonus- shepherd governor
6. Ensipazianna	6. Jared	6. Megalarus
7. Enmendurana	7. Enoch	7. Eueorachus
8. Enmeushumgalana	8. Methuselah	8. Amempsinus
9.Ubara-Tutu	9. Lamech	9. Otiartes
10. Ziusudra	*10. Noah*	*10. Xisuthrus*

Table 2: Table of Patriarchs Across Mesopotamia

PUTTING EVERYTHING TOGETHER

Chapters Six and Seven of the *Book of Enoch* tell how the nephilim, or race of giants, came to be. The angels (Watchers) saw the women of the children of men and lusted after them. There were two hundred angels that joined together in a pact to start families with human women (Enoch 6:6). The leader angel, named Semjâzâ, decided that he was going to find a wife and start a family. He did not want the other two angels to get in trouble and remarked:

"I fear you will not indeed agree to do this deed, and I alone will have to pay the penalty" - Enoch 6:3

However, the twenty leader angels informed Semjâzâ that they will all partake in a pact:

"Let us all swear an oath, and all bind ourselves by mutual imprecations not to abandon this plan but to do this thing" - Enoch 6:4[16]

. . .

In tablet nine of *The Lost Book of Enki*,[17] Sitchin helps to put all the pieces of this Sumerian, Babylonian, Akkadian, and Biblical story together of the Anunnaki, Igigi, Watchers, humans, and the flood. Tablet nine tells the story of Marduk's wedding to Sarpanit, the daughter of Enkime, and that two hundred Igigi came to attend. The leader *Igigi*, named Shamgaz, stated that he would bear the weight of the sin to spare the other Igigi. The other Igigi refused this gesture and said they would make an oath together. They took women and started families, calling the hybrid children "children of the rocket ships." Ninurta (Nimrod) went to the offspring of Ka (Ki) and taught music, alchemy, and how to build boats and cross the seas.[18] Genesis 6 gives a nod to the Igigi yet refers to them as the nephilim and states that they were on earth during the time of Noah:

"...and also afterward, when the sons of God came to the daughters of man, and they bore children to them. These were the mighty men of old, the men of renown."

– Genesis 6:4 NKJV

Enki and Ninmah (Ninhursag) visit the roof of Batanash. Enki takes Batanash and impregnates her. She gives birth to Ziusudra, making him a demi-god who obtained immortality per *The Epic of Gilgamesh*. The stories that appear around his time of birth are much like the stories that are told in Genesis before the birth of Noah and the birth of Atrahasis. Starvation, disease, and plagues were running rampant on the earth, and then, seemingly out of nowhere, the son of Enki was born. Ziusudra is raised by Batanash and her husband, Lu-Mach, in Shuruppak. Ninmah protects Ziusudra and Enki as Enki teaches Ziusudra the stories and writings of Adapa and priestly duties. Ziusudra marries Emzara and has three sons.[19] Genesis verse five says that Noah's father is Lamech, and Noah has three sons, Shem, Ham, and Japheth. The name of Noah's wife is not given. After the flood, in chapter ten, the narrator states that Ham becomes the father of the Canaan. Shem is the lineage that the

Hebrew scriptures follow, and Japheth's lineage travels to the region of Turkey.

If the purpose of the flood or great deluge was to get rid of the sinners and the nephilim, then why does Genesis 6:4 plainly state that the nephilim existed after the flood? If the nephilim were associated with sin, why were they called the "sons of God" and remembered as mighty men of renown? These are the titles and characteristics of a group that are highly regarded. When the *Book of Enoch*, *Lost Book of Enki*, and Genesis are all placed together, it shows that the Anunnaki came down, taught the children of men advanced knowledge and technology, and whose offspring were called the children of rocket ships.

All of the information in this chapter is known as ancient history. This means that the human race is not only here because of the extraterrestrial beings we look like but also because they are solely responsible for advancing our civilization in the time known as the Great Leap Forward. It is *only* in the first ten chapters of Genesis that the true story of mankind is told, whether it is summarized or not. After chapter ten, a very different story emerges as timelines are crossed or told in reverse. The retelling of Genesis chapter eleven and onward turns the plural "Elohim" into a singular "Yahweh," "Adonai," or as the church has named Him, "God."

Before the Eighteenth Dynasty

There is no definitive way of determining the timeframe between Enoch's life, the flood, and Abraham's life. However, the biblical stories have taken on a new dimension with the discovery of the Ugaritic texts. In 1928, near Ras Shamra, north Syria, a farmer's plow unearthed a discovery that had the potential to reshape our understanding of Christianity. An entire ancient Canaanite city named Ugarit was revealed along the coast of northern Syria. Within its ancient walls, the Temple of Ba'al and cuneiform tablets of stories, rituals, and commerce were found. This profound discovery shed light on the life of ancient Canaanites during the time of the 14th or 15th Dynasty of Hyksos in Egypt.

Josephus provides Manetho's records of the Hyksos (Hycsos) and their intriguing name: *Hyc[k]* (king), *sos* (shepherd). Manetho, an Egyptian by birth and a master of the Greek language, recorded Egyptian history in Greek. The shepherd kings, known as the Hyksos, held Egypt for a staggering 511 years before the Theban kings began their resistance. The Hyksos were eventually driven to a place called Avaris with ten thousand acres. After numerous conflicts, an agreement was reached and the Hyksos departed peacefully for the wilderness of Syria.[1] This period in

Egyptian and Canaanite history marked a significant shift that led to the rise of the 18th Egyptian Dynasty, a legacy that still echoes today.

The decline of the great Egyptian Pharaonic Dynasty can be traced back to the reign of Amenemhat I of the 12th Dynasty. His restructuring of the administration, which allowed each town to self-govern, led to a decline in the political influence of the pharaoh and Egypt as a whole. This trend continued during the reign of Pharaoh Khasekhemre Neferhotep I (1740-1729 BCE) of the 13th Dynasty, who did not have control over all of Egypt. This power vacuum allowed local regions within Egypt, such as Xios and Avaris, to increase their influence, setting the stage for the rise of the Hyksos and the subsequent 18th Egyptian Dynasty.[2]

The Hyksos people lived within the region of Avaris in Egypt. They were separate from Egyptians and were referred to as *Aamu*, later known as the Asiatics. As noted by Manetho, "Hyksos" came from the Greek. The Egyptian term for Hyksos was *hekau khasut*, or rulers of foreign countries, and was only used when speaking of the rulers of the Asiatics.[3] The Hyksos kingdom began during the 13th Egyptian Dynasty, causing overlap between the 13th and 14th Dynasties. Manetho gives the list of the rulers from the 14th to 17th Dynasties.[4]

The capital of the 14th and 15th Dynasties was located in Avaris (a four-square-kilometer region[5]) and controlled the Upper Kingdom (southern Egypt). This location was crucial because whoever controlled the apex of the Delta Nile controlled Egypt.[6] This caused the Egyptian pharaohs to rule from Thebes and to be cut off from major trade routes. Battles between Egypt and Avaris lasted two dynasties, and Hyksos built a fortified wall around Avaris. It was only at the end of the 17th Dynasty when Ahmose created a treaty with the king of Avaris to leave Egypt peacefully, causing a mass exodus of Habiru out of Egypt.[7]

An epithet from Nehesy[a], a high-ranking official in Avaris, possibly during the 14th Dynasty, gave praise to Seth instead of the normal praise given to Ra on Egyptian epithets. This recognition of Seth as lord

[a] Another name for Nubian (Shaw. *The Oxford History of Ancient Egypt*, p. 178.).

instead of Ra shows Seth was the patron deity of Avaris.[8] However, the following of Seth in Upper Egypt was not out of the ordinary. Sitchin notes in Egyptian mythology that when Ra divided Egypt into two, he gave Osiris Lower Egypt (the North), and Seth was given Upper Egypt. This led to the Osiris myth and caused the first Pyramid War.[9]

Archaeological evidence from this region shows a blending of cults from the area.[10] Yet texts from the Ugarit show that this may not be true. While there may have been a cult following in Avaris, the Ugarit texts show that the Canaanites wrote stories of Ba'al and had an entire pantheon of gods. The tablets found in Ugarit give a different view of the story that became the one god of Jewish religion.[11]

Instead of having only one god named Yahweh, the Canaanites of Ugarit had an entire pantheon of gods (Table 3)[12] whose traits and names mirrored those of other Sumerian and Babylonian pantheons. Many tablets containing the stories of the pantheon were destroyed with time, yet four stories remain: "Victorious Ba'al," "Ba'al Defeats Mot," "Legend of King Keret," and "Legend of Danel."

The "Legend of Danel" is similar to the story of Abraham and Sarah. Danel is unable to produce an heir to place a memory of him at Kadesh (spelled similarly to Qadesh, the Egyptian name of Jerusalem) and cries out to the gods. El and Ba'al arrive to tell him he will have an heir and to name him Aqhat ("The Pleasant One"). Aqhat obtains a bow created by El Kessem. Anat is jealous and promises Aqhat immortality in exchange for the bow. But Anat becomes irate after he tells her that the bow is made for warriors like him and not women. Anak asks El for permission to smite him and is told that she can punish him only up to a point. Anat entices Aqhat and takes him to the city of the "Lord of the Moon" and tells Taphan to kill Aqhat for his bow and then bring him back to life. Taphan kills him with three blows to the head, and before he can bring Aqhat back to life, his soul has left, and vultures are present to ravage the body. Ba'al assists Aqhat's sister in looking for his body while she finds Taphan and attempts to kill him in return.[13]

The overall theme of the "Legend of King Keret" is man's search for immortality, similar to *The Epic of Gilgamesh*. Keret's story is similar to

Job's; he loses his children, wives, and family through war and disease. His days and nights are filled with grief, and El descends to ask why he is grieving. The text then reveals that El is the father of Keret, making him a demi-god. He marries the daughter of King Udum, and Asherah blesses him and his wife by allowing the offspring to be semi-divine. The firstborn is named Yassib and is nursed by the goddesses Asherah and Anat. Keret's wealth grows, and he boasts of his divine place. Asherah becomes angry and sends a fatal disease upon him. Shataqat is able to save him, but Keret abdicates the throne, and Yassib becomes ruler.[14]

In "Victorious Ba'al," Hadad challenges Yam over the position of Ba'al. The Scriptural Research Institute notes that during the time they suggest these stories were written, approximately 1700 BCE, the main constellation of the vernal equinox changed from the Bull (Taurus) to the Ram (Aries). This caused the passing of kingship from El to Attar (the Ram).[15] While Yam was the original choice of El, he stated that Hadad had to fight Yam to declare the position. When he won, he made repeated statements about how he did not have a house or a temple. This lack of an official temple or residence could possibly show that the overall story had nothing to do with either a new god joining the pantheon or overturning the previously held beliefs and replacing it with a new god: a theme that is seen throughout the ancient world.[16]

The story of "Ba'al Defeats Mot" tells the events after "Victorious Ba'al." Ba'al is not satisfied, and with the help of Anat, they invite the other gods to a meal and kill them. Mot hears of this and challenges Ba'al over Earth's dominion. He accepts and when he arrives in the Lower World, he attempts to overthrow him but instead is killed by Mot. Anat learns what happened and makes the journey, ultimately slaying Mot. With the assistance of Shapash, they take Ba'al's body to the healers, and the god of magic brings him back to life.[17]

The names of the Canaanite pantheon appear all throughout the Old Testament. Asherah and references to her tree symbol, Ba'al, El, and Yam's stories, were part of the culture during the writing of the Old Testament. Chapter twenty dives deeper into the correlation between all the gods and discovers who is behind the Yahweh title. Before the story of Yahweh was told or the Roman Catholic Church created monotheism, the original Habiru were like other ancient cultures and believed in a pantheon of gods whose stories mirrored the Anunnaki of Sumer.

Name	Epithet
El/ Ab Adam	Father of gods, Father of men, The Kind One, Creator of all things, The Merciful.
Asherah	Consort of El
Yam	Lord of the Ocean and Sea Beloved of El
Haddu/Hadad	Storm God
Ba'al	Lord of Zaphon Cloud-Rider
Mot	The Smiter, Annihilator God of Death Son of El
Anat	She Who Responded Sister and lover of Ba'al The Virgin
Shataqat/Shepesh	One who removes illness Goddess who knows all magic Torch of the Gods
Kothar-wa-khasis	The skilled and knowing The god of metalsmithing
Lotan	The Seven-Headed serpent
Atak	The Bullock
Hashat	Star B!t*h – referring to Isis and Sirius
El Kessem	The Craftsman of the gods The god of Magic
Attar	Ram God Possessor of Kingship
Elat	Goddess of the Underworld

Table 3: Ugaritic Pantheon

The Eighteenth Egyptian Dynasty

The first ten chapters of Genesis summarize the story of real beings who came to Earth from the stars and lived among people. Whether or not these beings were still present when the stories began is unknown. The question is, what happened to all the other Elohim? Did they disappear, or was there a story so crazy that it had to be subverted, clipped, edited, and retold to forget what happened and who caused it? Understanding the importance of the 18th Egyptian Dynasty changes how the Old Testament is read. It fundamentally changes the entire story of Yahweh and his people and answers this question. Chapter ten of Genesis ends with Noah's genealogy, and a chapter later, the story of Abram begins.

Take the time to read the Old Testament and discover that the story does not make sense. Genealogies, timelines of events, locations, and weird terms are used to hide a story lurking under the surface. Most churches thrive because people do not read the entirety of the Old Testament for themselves and rely on what the teacher says. The biblical cover-up is what it should be called, and the patriarchs (Abraham, Isaac, Jacob, Joshua, Moses, King David, and King Solomon) are reiterations of people from history with changed timelines. This reiteration may seem minor or insignificant, but it has a lasting impact on humanity's worldview. Extraterrestrial contact knowledge in the Bible is no small matter, much less numerous beings who masqueraded as gods because of their knowledge of advanced technology.

Egyptian history was written at the time or within a century of an event. This means that information was accurately depicted on reliefs from temple walls and monuments. Ugarit texts show that the history of the Israelites is now what modern Christianity claims as truth. As the game of telephone goes, information gets lost and mistranslated, whether accidentally or intentionally. Flavius Josephus remarks on the accuracy of Egyptian history. He notes that the Egyptians took special care in providing accurate information because history was sacred to the culture and was shown publicly.[1]

There is no correlation between the 18th Dynastic Pharaohs and the biblical patriarchs for good reason. This dynasty singlehandedly caused the most influential change in Egypt's cultural and religious history. This forever changed how the gods were revered in Egypt and the rest of the ancient world. This change caused the Amarna kings to be left out of the Abydos Kings List. To preserve the memory, the biblical storytellers changed the names from Egyptian to Hebrew and weaved multiple stories together. This skewed the original stories and created fantastic fiction for a new religion to believe. Continuing to understand the Egyptian Dynasties in relation to the Ugaritic text and pre-Jewish beliefs shows this transformation.

The Eighteenth Dynasty

The rules of succession in Egypt were very strict and followed the rules laid out by the Egyptian gods. Ra ruled over Egypt for a thousand years (more background information will be given in chapter twenty). When Ra decided to leave Egypt, he handed rule to his children Shu and Tefnut. They, in turn, passed succession to their children Geb and Nut, who passed succession to their four children: Asar (known as Osiris to the Greeks), his half-sister-wife Ast (known as Isis, fathered by Thoth), Seth, and his sister-wife Nephthys. These four became the true gods of Egypt because the Egyptians believed the gods were born out of the union of heaven and earth.

It was Egyptian royal custom that a god, and therefore man, could have multiple concubines and affairs that produced children. This made the first rule of succession significant. The heir was the first son born to the pharaoh's half-sister (official spouse or Great Royal Wife). If she could not have a son, succession moved to the first concubine or affair that gave a son. Yet if, at any time, the Great Royal Wife gave birth to a son, he would become the heir.[1]

Ahmose or Amosis I

Ahmose I was the son of Pharaoh Seqenenre Tao and brother to Kamose, the last pharaoh of the 17th Dynasty. Ahmose spent the first part of his reign fighting against the invading Hyksos (Semitic shepherds). Josephus details this victory and that it was won with a treaty. He also breaks down Hyksos/Hycsos as *Hyc,* meaning 'king,' and *Sos,* meaning 'shepherd.'[a,2] This act caused Ahmose to establish himself as Ahmose I and began the 18th Dynasty.[3]

Ahmose I also seized a stronghold in Southern Palestine at Sharuhen and began a military campaign in Kush's capital of Kerma.[4] His rule was characterized by the unification of Egypt and regaining land previously lost during other dynasties. After obtaining previously lost land from the Hyksos, he began constructing a commercial center for government use in Avaris.[5] While his name meant "son of the moon god, Iah,"[b] he contributed to the Cult of Amun at Karnak. He began restoration work at the Karnak Temple of Amun and wanted to be recognized as a devoutly dedicated king to Amun.[6] However, this was just his outward appearance. He was part of the elite Brotherhood that began during the reign of Khufu. He held mystery school classes in his private chambers.[7]

Amenhotep I

Ahmose I's elder brother Ahmose-Ankh was the original heir to the throne but died before Ahmose I. Amenhotep was second in line for secession. He successfully completed military campaigns begun by Ahmose I in Kush[c] and brought in increased financial wealth that improved the economy. He reigned alongside his mother, Ahmose-Nefertari, after Ahmose I's death and was worshiped and deified in Thebes. Further military conquests in Nubia extended Egypt's border

---

[a] Josephus also accounts for the reason that *Hyc* is associated with the translation as captive and notes that it originates back to Joseph who told the king of Egypt that he was captive (Josephus. Apion 1:83, 92).

[b] The breakdown of Ahmose's shows the meaning: *Ah* referencing Iah, the moon-god, and *mos* meaning son or child (Bryan 209).

[c] Breasted notes that Amenhotep was still under Ahmose I when the Kush campaign ended (Breasted 17).

south, which helped to guarantee relations with powerful families such as Elkab, Edfu, and Thebes. This, in turn, helped to create an administration organization in the region.

Twelve years of peaceful rule followed after a minor rebellion in Nubia. This allowed Amenhotep to open mines, continue building monuments, and rebuild the Karnak temple of Amun. One of the most important characteristics of Amenhotep's reign, which would become the hallmark of the 18th Dynasty, was devotion to Amun's cult at Karnak.[8] Yet this devotion helped to continue the outward devotion to Amun while attending mystery school teachings. He even became a teacher of the mysteries.[9]

Thutmose I

Thutmose I was a general of Amenhotep I and not an official member of the Ahmosid family by blood.[10] He became pharaoh after Amenhotep I arranged the marriage between Thutmose I and Ahmose (daughter of Ahmose-Ankh, brother of Amenhotep I) and appointed him coregent.[11] During his short reign, he established himself and made Egypt superior over every land[12] by beginning military expeditions into Syria that allowed for later diplomacy. He led military campaigns eastward through Nubia and Kush's capital in Kerma to Syria and restored the city of Niy.

His major change, however, was that he promoted the connection between god and king along with the connections between kings. Because he married into the royal family, he began associating with kings of the past over the present ones.[13] This act by Thutmose I began to separate current knowledge skew history to create a specific narrative in the minds of Egyptians.

A stele at the temple of Osiris in Abydos shows Thutmose I professing to the priests that he was born of Osiris and sent by the earth-god Geb and Ptah to use his vast wealth and restore the sanctuaries of the gods. It details his speech of universal triumph and refers to himself as the favorite of Amon, a son of Re. He even positions himself alongside the gods and calls himself Lord of Abydos, Ruler of Eternity who will shine as King forever, as Re rules forever.[14] This continued the message of separation from previous kings and the rightful king of Egypt, the son of Re. His wife was a member of the Brotherhood and was equal with the men in the society.[10] Their work with the Brotherhood helped create the circumstances surrounding the end of the 18th Dynasty and the beginning of the Hebrew stories.

Thutmose II

Son of Thutmose I and his minor wife Mutnofret, he was not heir to the throne. He married his half-sister Hatshepsut, the daughter of Thutmose I and Queen Ahmose. He ruled after the death of Thutmose I, and his years of sole rule were brief, possibly no more than three years. A stele south of Aswan showed a rebellion in Kush, the capture of prisoners, including the king's children, and the death of every male but the king's son. This massacre ended with continuous problems within the region.[15] Thutmose II was the father of Thutmose III by a concubine named Isis. To safeguard his son's right to rule, he took Thutmose III to the Temple of Amun in Karnak to be adopted by the state god Amun-Re. Thutmose II's attempt was thwarted after his death by his wife and half-sister Hatshepsut.

Hatshepsut and Thutmose III

Hatshepsut

Hatshepsut was the daughter of Thutmose I and Ahmose, and was the Great Royal Wife to Thutmose II. She took on the double crown of Egypt—King of Upper and Lower Egypt—shortly after the death of Thutmose II. Hatshepsut stepped in and executed the affairs of Egypt and ruled for eighteen years. Reliefs show Hatshepsut as a woman during the beginning of her reign, yet she became depicted as a man and king towards the end.[16] There were a few military expeditions during her reign, but it was marked by a time of peace and prosperity.

Breaking from the tradition set forth by Thutmose I, she used her ancestral connection to previous rulers through her mother's bloodline (Ahmose and Ahmose-Nefertari) to legitimize her right to rule. While still using the title of "God's wife of Amun," Hatshepsut attempted to return to the previous rules of succession set forth by the gods.[17] Her only known offspring was Neferure, and she attempted to prepare her daughter for her role as heiress. Neferure died in year six of her coregency, and Hatshepsut again placed Thutmose III in the background.[18]

Egyptian History shows a frequent pattern of pharaohs erasing the names of their forebears, often restoring them later. This was especially true during the reign of Hatshepsut. She erased the names of her two predecessors and placed her own. When Thutmose III solely ruled over Egypt, he erased her name and returned the names of Thutmose I and II to their rightful places.[19] Bryan remarks that Hatshepsut's name was systematically removed from monuments attributed to her. Thutmose III wanted to eliminate her family line and move away from the Egyptian succession rules and towards being chosen by the state god Amun-Re.[20]

Thutmose III

Thutmose III was given a wide range of ages when he was adopted by the state god Amun-Re, but the adoption is indisputable and is shown on a pillar at Karnak:

139. "His majesty [Thutmose III] placed for him incense upon the fire, and offered him a great oblation consisting of oxen, calves, mountain goats.

140. On recognizing me, lo, [Amun-Re] halted ---- [I threw myself on] the pavement, I prostrated myself in his presence. He set me before his majesty...

141. [He opened for] me the doors of heaven; he opened the portals of the horizon of Rd. I flew to heaven as a divine hawk, beholding his form in heaven; I adorned his majesty... feast. I saw the glorious forms of the Horizon-God upon his mysterious ways in heaven.

142. Re himself established me. I was dignified with the diadems which [we]re upon his head, his serpent-diadem, rested upon [my forehead] ... [he satisfied] me with all his glories; I was sated with the counsels of the gods, like Horus, when he counted his body at the house of my father, Amon-Re. I was [present]ed with the dignities of a god, with... my diadems.

143. His own titulary was affixed for me.

146. ... [in this my name]. King of Upper and Lower Egypt, Lord of the Two Lands: "Menkheperre" (the being of Re abides)."

Chapter fifteen will go into more detail about the reign of Thutmose III. He was not the son of the Great Royal Wife, he did not marry her daughter, and he was not a descendant of the Ahmoside Dynasty. This means that technically, in the eyes of Egyptian succession, Thutmose III had no legal right to the throne. *His right to rule was solely based on the fact that he was chosen by the state god Amun-Re.* This movement, which moved away from Egyptian succession rules to being chosen as ruler by Amon-Re, was founded by Thutmose III and would be followed until the end of Aye's rule.[21]

Yet the Brotherhood was behind his rise to pharaoh. The chief priest of the Brotherhood and temple said that Thutmose III had been chosen by Ra instead of his brother. Thutmose III singlehandedly helped shape the Brotherhood into what it is today. He projected the outward picture of devotion to Amun due to political conditions while successfully creating a secret organization. Like the pharaohs that came before him, Thutmose III held secret mystery school classes in his chambers. To avoid outside voices, the council became completely closed and thus started the Grand Council Meetings. The council became known as the Brotherhood and only acted in secret, and any correspondence contained the statement relating to the last circle within the order, the Illuminati. He was also responsible for defining the rules, procedures, and principles of the Brotherhood, which have been unchanged since. These procedures and principles included plans to begin movement north into Europe and west into the Americas.[22]

. . .

Amenhotep II

Born to Great Royal Wife Merytra and Thutmose III, Amenhotep II's reign as pharaoh is often mistakenly overshadowed by his predecessors. His reign was mostly peaceful and pivotal to the flourishment of the 18th Dynasty. While there was a revolt from Syria after the death of Thutmose III,[23] Amenhotep II formed an alliance with Naharin (Mitanni) that had not previously been obtained by his predecessors.[24] Unlike his father, who focused on military strength, Amenhotep II's reign focused on cultural improvement by allowing agricultural, industrial, and administrative organizations to flourish. This change within the culture shaped even the art created and began to show artistic individualism.[25]

He continued his father's building projects at Karnak. He paid homage to the other Egyptian gods at various temples but focused most of his time on buildings dedicated to Amun-Re. A Steele at Karnak notes his kingship given to him by Amun-Re, with the exact wording used as that of Thutmose III.[26]

Thutmose IV

Thutmose IV was the son of Amenhotep II and possibly his wife Tiaa. Her position within the royal family is unknown, but Thutmose IV perhaps placed her on a monument with Amenhotep II during his reign. The Egyptian rights of succession are not known for Thutmose IV, but the Dream Stele found at the Sphinx showed that the historical rules of succession did not matter, and the 18th Dynasty pharaohs ruled because they were chosen by the sun god Amun.

The Sphinx Stela, better known as the Dream Stele, tells the story of Thutmose IV's vision when he was young. It begins with Thutmose IV resting in the shadow of the great god around midday. A vision occurred, and he "found the majesty of this revered god speaking with his own mouth, as a father speaks with his son, saying, 'Behold thou me!

See thou me! My son Thutmose. I am thy father, Harmakhis-Khepri-Ra-Atum, who will give to thee my kingdom on earth at the head of the living. Thou shalt wear the white crown and the red crown upon the throne of Keb, the hereditary price.'"[27] Breasted notes that this stele is not written in the style of official or royal records.

The problem of succession based on the birth order to the correct royal wife no longer mattered because of the new rules of kingship established by his predecessors. He used his building campaigns to cement his relationship with the sun god. He focused his devotion on Horemakhet-Khepri-Ra-Atum (the northern sun deity) and the Heliopolitan cult. There are no references to Amun-Ra found on the Sphinx Stele, which could have been deliberate as Heliopolis increased its importance within political and administration influence.[28] While Thutmose IV sided with the northern sun god in Heliopolis, at Karnak, his inscriptions leave out the north sun deity, only speaking of Amun-Ra.

At Karnak, Thutmose IV completed a single obelisk originally produced for Thutmose III before his death. This is known as the Lateran Obelisk. Thutmose IV made Thutmose III's inscription in the middle lines, while Thutmose IV wrote his inscriptions along the right and left sides of his grandfather's inscriptions.[29] The inscriptions that Thutmose IV wrote are similar to inscriptions made by his predecessors: "Thutmose IV, begotten of Re, beloved of Amon."[30] He deliberately aligned himself with the sun god of both north and south Egypt and emphasized the divinity of royal females. While his Great Royal Wife was his mother, Tiaa, he married Artatama, daughter of the Mitanni Ruler, to continue a diplomatic relationship with Mitanni.[31]

There is a stele of Thutmose IV giving an offering of milk to Hathor with a man behind him holding a loaf of bread. The inscription says this man is named Neby, the Mayor of Tjaru. This story of Neby is the first time any references are made to Aten in the 18th Dynasty. He traveled with Thutmose IV to Sinai and was commissioned in Queen Tiaa's home. Björkman notes that while attempting to translate a mysterious inscription associated with Neby (*imy-r hnu*), Professor Jean Yoyotte showed him a photograph from the Pierre Lacau Archives (Photo A III, 63, F6). This photograph was of a small block (*talatat*) from Akhenaten's period. It revealed the translation to be "The Overseer of the Foremost Water in the *hunt* (watercourse/canal) of the Temple of Aten."[32]

The land bridge connecting Egypt with the Middle East is known to the Egyptians as the Way of Horus, and is a natural route for all types of travel. The Hebrews called this land bridge the *Way of the Land of the Philistines* (Exodus 13:17). The town of Tjaru, also known as Zarw or Zerukha, is a fortress along this land bridge. To the Hebrew people, this town was known as Goshen,[33] where the story of Moses begins.

Chronology of the Eighteenth Dynasty[a]

18th Dynasty Pharaohs	Great Royal Wife & various titles	Speculative Dates BC[b]
Ahmose I Nebpehtire[c]	Ahmose-Nefretiri Great Royal Wife God's wife of Amun	1550-1525
Amenhotep I Zeserkere	(Ahmose-) Merrtamun[d] Great Royal Wife United to the white crown Lady of the Two Lands	1525-1504
Thutmose I Okheperkere	Ahmose Great Royal Wife King's Sister[e]	1504-1492
Thutmose II Okheperkere	Hatshepsut King's Daughter King's Sister Great Royal Wife God's wife of Amun	1492-1479
Hatshepsut Maatkara		1473-1458
Thutmose III Menkheper-Ra	Sitiah Great Royal Wife God's wife of Amun[f]	1479-1426
Amenhotep II Okheprure	Meryta Great Royal Wife God's wife of Amun Mother of Amenhotep II	1427-1400
Thutmose IV Menkheprure	Tiaa[g] King's Mother God's wife of Amun	1400-1390

	Great Royal Wife	
Amenhotep III Nibmare	Tiye Great Royal Wife	1390-1352
Amenhotep IV Akhenaten Neferkheprure-Wanre[h]	Nefertiti Neferneferuaten	1352-1336
Smenkhkare	Meritaten Great Royal Wife Daughter if Nefertiti	1350-1347
Tutankhamun	Ankhesenpaaten Ankhesenamun[i]	1336-1327
Aye		1327-1323
Horemheb Zeserkheprure		1323-1295

Table 4: Chronology of the Eighteenth Dynasty

Chronology of the 18th Dynasty

[a]18th Dynasty Pharaohs Great Royal Wife & various titles
Speculative Dates BC

Ahmose I
Nebpehtyre[b]
Ahmose-Nefertari
Great Royal Wife
God's wife of Amun
1550-1525

Amenhotep I
Zeserkere

---

a These dates are very approximate and change depending on source material. Approximate dates are given based on *The Oxford History of Ancient Egypt.*

b Egyptian names were given at birth. Once they rose to the position of Pharaoh, they took on the name they are commonly known by. All sources of the Pharaoh's Egyptian name were obtained from Breasted (*Ancient Records of Egypt Volume II: The Eighteenth Dynasty*).

(Ahmose-) Meritamun[c]
Great Royal Wife
United to the white crown
Lady of the Two Lands
1525-1504

Thutmose I
Okheperkere
Ahmose
Great Royal Wife
King's Sister[d]
1504-1492

Thutmose II
Okheperkere
Hatshepsut
King's Daughter
King's Sister
Great Royal Wife
God's wife of Amun
1492-1479

Hatshepsut
Maatkare
1473-1458

Thutmose III
Menkheper-Ra

---

[c] Bryan notes that (Ahmose-) Meritamun as Amenhotep I's wife is speculative. No supporting documentation is found, and it's speculated that her coffin was stylistically similar to those found during his reign. There is a speculative reference to her in Amenhotep's year 8 stele in Nubia (Bryan 219-20).

[d] Daughter of Ahmose-Ankh, brother to Amenhotep I who was supposed to inherit the throne but died before Ahmose I died allowing Amenhotep I to take his place.

Sitiah
Great Royal Wife
God's wife of Amun[e]
1479-1426

Amenhotep II
Okheprure
Meryta
Great Royal Wife
God's wife of Amun
Mother of Amenhotep II
1427-1400

Thutmose IV
Menkheprure
Tiaa[f]
King's Mother
God's wife of Amun
Great Royal Wife
1400-1390

Amenhotep III
Nibmare
Tiye
Great Royal Wife
1390-1352

Amenhotep IV
Akhenaten

---

[e] Royal titles are only seen in one surviving text. Sitiah was the daughter of a royal nurse and was replaced by Merytra. The Temple of Medinet Habu and Thutmose III's tomb showed Merytra as queen to Thutmose III (Shaw. *The Oxford History of Ancient Egypt,* p. 240-41.).

[f] Amenhotep II's jubilee pavilion shows Meryta changed her name to Tiaa. It is not known if these two were the same person.

Neferkheprure-Wanre[g]
Nefertiti
Neferneferuaten
1352-1336

Smenkhkare
Meritaten
Great Royal Wife
Daughter if Nefertiti
1350-1347

Tutankhamun
Ankhesenpaaten
Ankhesenamun[h]
1336-1327

Aye
1327-1323

Horemheb
Zeserkheprure
1323-1295

---

[g] Birth name changed to throne name, Amenhotep IV. He changed his name to Akhenaten shortly after in reverence to Aten.

[h] She changed her name to honor previous Egyptian deities.

The Amarna Kings

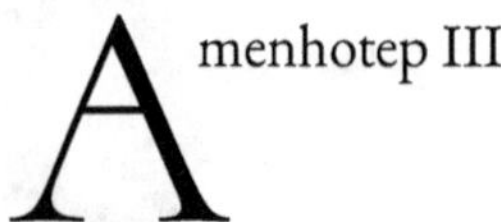

Amenhotep III

Amenhotep III's reign is classified as the pinnacle of the 18th Dynasty, and under his reign, this pinnacle was obtained without military achievement.[1]

The Egyptian pharaohs are always thought of as looking Egyptian, yet the busts of Amenhotep III and Queen Tiye show negroid features. The busts of the pharaoh and his queen show that they are of negroid descent and not the previously thought Egyptian descent.

Figure 12: Amenhotep III

Figure 13: Bust of Queen Tiye
Credit: Dosseman CC: BY-SA 4.0, Wiki Commons

Chapter sixteen expands upon the reign of Amenhotep III and the possible correlation with Solomon. In this chapter, it is essential to point out the trend of this dynasty. He married his half-sister Sitamun when she was young, but his wife Tiye held the title of Great Royal Wife. Amenhotep III and Queen Tiye were deified during their lifetimes. He unified Egypt, and there was no longer a North and South Egypt; it became one land instead of two. At the time of his first jubilee, Amenhotep III intended to be identified with the sun god Ra. Festival paintings show Amenhotep III riding in the solar boat and taking the place of Ra.

Monuments were erected that showed Amenhotep III merging with the sun god before death, changing the previous Egyptian custom and belief that pharaohs merged with the gods after death. The most important aspect of his reign was the transformation of the Egyptian pantheon of the gods. His building projects at Karnak stressed royal identification with the sun god Amun-Re. He changed the previous orientation of the

Karnak Temple to represent the solar eye of Ra. This exhibited that he could create stability within the cosmos, an action that further aligned himself with Ra.[2]

Diplomacy and international relations with foreign people grew, and people began to see themselves as a part of god's creation and not separate from it. Ra's rule then sustained them, and the chosen pharaoh was Ra's representative on Earth. Cults across Egypt became solarized, and instead of having many gods, all gods were merged into different aspects of Ra. The daily ritual of sunrise to sunset represented Ra's various forms. At night, Ra would die and be reborn, making Osiris the god of the underworld, an aspect of Ra. Instead of the ancient teachings of the gods, there was only one god, Ra, the primeval creator. All other gods were created from him; therefore, all gods were different aspects of him.[3]

Along with merging all the gods of Egypt into aspects of the one god Ra, he named his Theban palace "the gleaming Aten." Up until the reign of Amenhotep III, Memphis was the location of the primary royal residence. Amenhotep III built his main royal residence in Thebes, the religious and administrative capital. He built a pleasure lake at his home in Thebes with a commemorative scarab detailing the lake's construction for Queen Tiye.

"His Majesty commanded to make a lake for the Great King's-Wife, Tiye, in her city of Zerukha (Zarw). Its length is 3,700 cubits; its width, 700 cubits. His majesty celebrated the feast of the opening of the lake, in the third month of the first season, day 16 when his majesty sailed thereon in the royal barge: 'Aton[en]-Gleams'."[4]

This scarab and the naming of Amenhotep III's palace show that while Aton (Aten) was not widely circulated in Egypt, the pharaohs seemed to already have a plan in motion that would transform all of Egyptian society and have lasting effects on the world. It also showed that the influence of the Brotherhood had continued after the death of Thutmose III. He was known as a Master of the Order within the Brotherhood and after the Temple of Luxor was complete, he dedicated it to the Order.[5]

Amenhotep IV/Akhenaten

Osman uses the Pleasure Lake scarab, the Old Testament, and other Egyptian sources to deduce that Akhenaten was born in Zarw near the biblical city of Goshen along the eastern Delta.[6] Akhenaten was the second son of Amenhotep III and Queen Tiye and not the officially crowned prince. His older brother, Thutmose, was crown prince but died prematurely.[7] Chapter twelve describes in detail the similarities between Akhenaten and Moses.

Akhenaten's upbringing in Zarw could have played a significant role in the changes he would eventually make. Queen Tiye's brother, Anen, was a high priest of Re at Heliopolis and could have trained and educated Akhenaten during his years before coregency. Elements of solar worship from Heliopolis are found in Atenism and could be linked to his uncle. Osman remarks that if Akhenaten had been brought up in Zarw, he would have grown up learning about the ideas of an imageless god.[8] He was instructed in the secret mysteries of the Brotherhood as a child and when he was 13 years old, he was declared a Master by Council Decree.[9]

The religious changes that occurred during the reign of Akhenaten changed history. Akhenaten used his predecessor's example and began his reign by professing that Amun had chosen and crowned him. He started building constructions at the Karnak Temple of Amun and made his first controversial statement. Instead of dedicating his new temple to Amun, he dedicated it to a new form of the sun god: "The living one, Ra-Horus of the Horizon, which is in the sun disk." He changed the sun god to the living sun-disk (shortened to sun-disk) or Aten. Amenhotep III brought about the worship of the visible sun, and Akhenaten began worshiping the sun's rays that touched the royal family. This change took place around the second to third year of his reign.[10] He believed this change was necessary because the people of

Egypt worshipped the sun as a god instead of as a *symbol* of god, marking the beginning of monotheism in Egypt.[11]

Figure 14: Akhenaten

By the fifth year of Akhenaten's reign, he severed all ties to the religious capital in Thebes. He changed his name from Amenhotep IV to Akhenaten and created a new capital for the cult of Aten, called Akhenaten (*Horizon of the Aten*), known as El Amarna today.[12] Osman notes a variance in this history and states that Amenhotep built a temple to the Aten in Thebes alongside the Amun-Re temples within his first year as coregent with Amenhotep III.[13] The new capital at El Amarna marked the beginning of the Brotherhood, not associated with the pharaoh. This new phase showed that the Brotherhood had become so restrictive that even the priesthood was not allowed into the Order, [14] making it a concrete organization.[15]

Queen Tiye ensured her son was heir to the throne through an arranged marriage to his half-sister Nefertiti, daughter of Amenhotep III, and his sister Sitamun, to satisfy Egyptian custom upheld by Egyptian priests and nobility.[16]

Back in Karnak, Akhenaten elevated his Great Royal Wife, Nefertiti, to a status equal to his own. Before the reign of Akhenaten, this had never been done. Her name was changed from Nefertiti to Neferneferuaten, and a relief at Karnak shows her alone or with her daughter Meritaten performing the rights reserved by the king. Instead of Nefertiti acting as co-regent with Akhenaten, they ruled together.

This change from co-regent to ruling side by side fundamentally changed the belief structure held in the traditional triad of Atum as the primeval father and the twins Shu (son) and Tefnut (daughter). This dual rule replaced the triad to Father Aten with the living king and queen, earthly representatives, and children of the Aten. This change embedded Akhenaten's status and the father-son relationship between Aten and the pharaoh. Osman notes that Akhenaten wanted to be seen as Shu because he rejoiced in the horizon and was a god represented in human form.[17]

Figure 15: Queen Nefertiti

Credit: Miguel Hermoso Cuesta.

After the death of Amenhotep III, during the twelfth year of Akhenaten's coregency, temples of Egyptian gods were shut down, funding was diverted to the cult of Aten or cut off, priests were left to return home, religious festivals, processions, and holidays honoring the traditional Egyptian gods stopped.[18] Whereas the previous Egyptian custom held that the pharaoh was the head of the priesthood, temples, army, and administration, Akhenaten separated himself from the status of the temples and priesthood. He relied heavily upon the army to avoid possible confrontation with the people of Egypt. The lasting effects of Akhenaten's transformation are seen in Tutankhamun's Restoration Stele:

"Now when His Majesty arose as king, the temples of the gods and goddesses, beginning from the Elephantine [down] to the marshes of the Delta, [their? ----- had] fallen into neglect, their shrines had fallen into desolation and become tracts overgrown with plants. Their sanctuaries were as if they had never been,

their halls were a trodden path. The land was in confusion, the gods forsook this land. If an [army? was] sent to Djahy to widen the frontiers of Egypt, it met with no success at all. If one prayed to a god to ask things of him, [in no wide] did he come. If one made supplication to a goddess in like manner, in no wise did she come. Their hearts were weak of themselves (with anger); they destroyed what had been done."[19]

He was intent on removing all references to Amun or Ammon and even removed the names of those from the 18th Dynasty because their names were placed to appease the priesthood, and all other forms of worship were banned.[20] In addition to all the other changes Akhenaten made, he implemented a system that incorporated all the teachings of the Brotherhood into a code of secret symbols, making true knowledge unattainable to everyone except the Brotherhood.[21]

Nefertiti and Akhenaten had six daughters (Figures 16 and 17 show three daughters), and it is speculated that the mother of Tutankhamun was the Mitannian princess, Kiya Akhenaten's second wife, but it is unknown.[22]

It is interesting to note the various ways that Akhenaten is seen by many throughout the millennia. Charles F. Potter notes that he was the first pacifist, realist, monotheist, heretic, humanitarian, and first known to attempt founding a new religion that shone like a comet and then left. He was born thousands of years too soon but helped begin an intellectual evolution.[23] Manly P. Hall writes of him in a positive light, a universal god shining through all creation. He was also the first pharaoh in history to choose to portray himself with his arm around his wife with his seven daughters present.[24]

Smenkhkare

The grand flip was achieved, and the aftermath was felt for centuries in Egypt. The throne was in disarray, causing Akhenaten's brother Smenkhkare to become his co-regent. Akhenaten was informed by his uncle Aye that there was a threat to his life, and he fled to Sinai after abdication.[25] Shaw notes that the identity of Smenkhkare is unknown and speculates that Nefertiti could have disguised herself as a man, similar to Hatshepsut, and ruled until the possible death or fleeing of Akhenaten.[26] Meritaten (Figure 16), daughter of Akhenaten and Nefertiti, served as Great Royal Wife to the mysterious Smenkhkare. Little is known of this pharaoh, and Tutankhamun replaced them after three years.

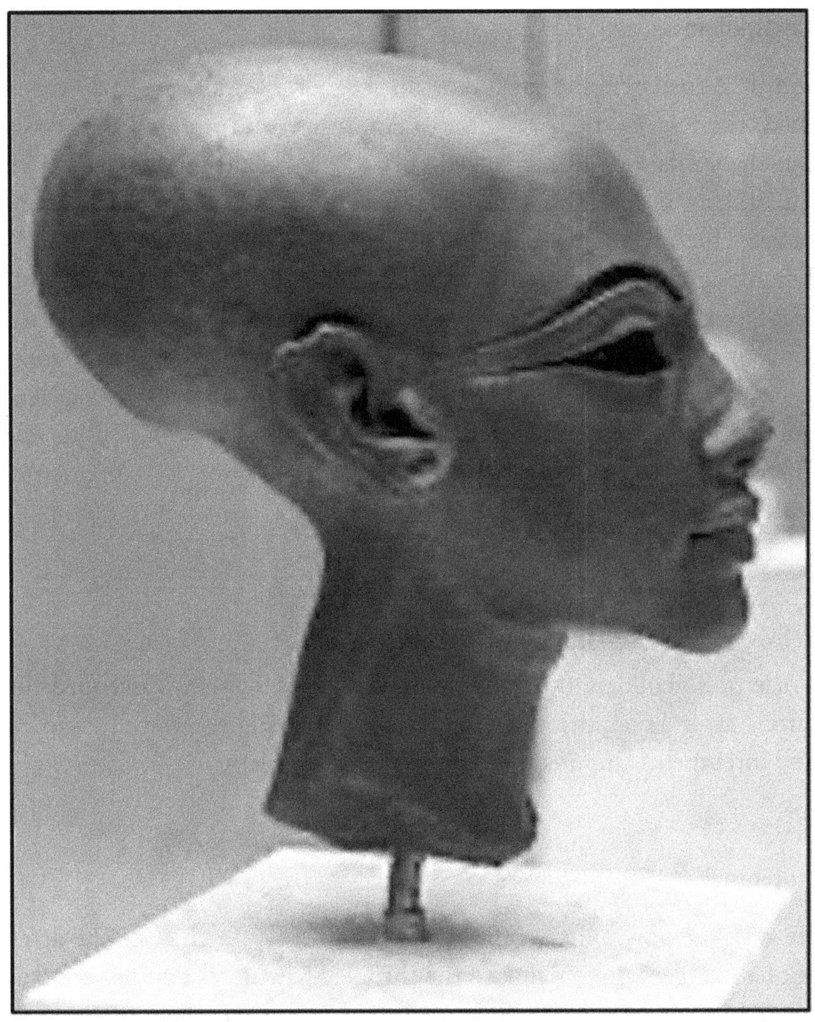

Figure 16: Meritaten Eldest Daughter of Akhenaten and Nefertiti

Credit: Miguel Hermoso Cuesta CC:

Tutankhamun

Son of Akhenaten and an unknown mother, he took the throne as a child, restored Egyptian religion to Amun-Re, and restored previously demolished sites. He changed his birth name Tutankhaten to Tutankhamun, and his Great Royal Wife and half-sister Ankhesen-paaten followed suit, changing her name to Ankhesenamun.[27] Tutankhamun ruled for only nine years before he died. The cause of his death is unknown, and modern science speculates that a myriad of congenital deformities and complications produced from incest within the royal family ultimately led to his early death. Other theories postulate that he possessed non-terrestrial DNA.[28] This caused him to be not fully human and not entirely extraterrestrial, leading to a premature death.

Aye

Uncle of Tutankhamun and brother to Queen Tiye, Aye ascended the throne after Tutankhamun's premature death. He successfully warded off Horemheb's attempt at the throne until his death.

Horemheb

He was a senior military official and commander-in-chief of the army but had no genealogical link to the throne. Shaw notes that he served as a stand-in coregent to Tutankhamun.[29] Once Horemheb ascended to the throne, he sought to destroy all traces of the Amarna kings and their legacy. He openly declared his right to rule at the Opet Festival procession with a public appearance of Amun through an oracle.[30] He proudly professed his detest towards the Amarna kings and their followers with the Vienna Fragment.

"Asiatics; others have been placed in their abodes----- they have been destroyed, and their town laid waste, and fire has been thrown ----; [they have come to entreat] the Great in Strength to send his mighty sword before -----. Their countries are starving, they live like goats of the mountain, [their] children --- saying" 'A few of the Asiatics, who knew not how they should live, have come [begg]ing [a home in the domain] of Pharaoh after the manner of your fathers' father since the beginning, under----.'"[31]

The Lasting Effects

The 18th Dynasty shows that ideas and worship can be changed throughout a century. It is a gradual process through which the powerful few change people's minds. Ahmose I focused on family relationships and the cult of Amun. Amenhotep I was devoted to the cult of Amun and added to the Karnak temple of Amun. Thutmose I was not born into the royal family but married into it. He created the connection between god and king before king and king, and positioned himself alongside Re. Thutmose II aligned with the Cult of Amun and ensured that Amun-Re adopted his son before his death. Hatshepsut attempted to return to the previous familial ties but was thwarted by Thutmose III. He openly stated that Amun-Re adopted him and that his right to rule came from being chosen by Amun-Re. Amenhotep II focused on the cult of Amun and expressed that his kingship was given by Amun-Re. Thutmose IV was not the heir to the throne but wrote the Dream Stele that showed Re chose him. He gave devotion to both North and South Egyptian sun god cults. Amenhotep III united Egypt, rearranged the Karnak temple, built the Luxor temple, and showed that all other Egyptian gods were only aspects of the one true god, Ra.

Figure 17: Neferneferuaten Tasherit and Neferneferure. Daughters of Akhenaten and Nefertiti

This allowed Akhenaten's emergence and monotheism's beginning through Ra to flourish. The artistic style of the Amarna period was changed from a formal depiction to a more realistic representation of the royal house. There are two sides to the argument of how the royal family was depicted:

1. It was an exaggerated caricature, or

2. It was an accurate representation.

Remembering the presence of elongated skulls from Chapter two, this family could have been a part of the ties to pre-diluvian societies or gods. While the truth is unknown, speculation tied with factual evidence presented in this book points to the second. This theory leads to the idea that there was a plot toward monotheistic control that the Catholic church ultimately dominated.

The Biblical Patriarchs: Abraham to Joseph

When the study of the Old Testament for this book began, the original idea was to show evidence of extraterrestrial contact and advanced knowledge and technology in the Bible. What was found instead was an entire rewording of history changing the entire story of Christianity. Dr. Alan Snow remarks that the Hebrew within the Bible was not developed until the time of Solomon. The first five books of the Bible were not even written until the time of Rehoboam. Chapter eight showed that the Habiru or Hyksos lived in and around Egypt. The Habiru used the stories they were familiar with—the pharaoh's lives and wars—and placed their own spin on it. This means that the stories of Abraham to Solomon are the stories of pharaohs.[1] Through decades of research, Ahmed Osman uncovered this truth and found the link between the biblical patriarchs and the Egyptian pharaohs of the 18th Dynasty.

Wait, what do the Egyptian pharaohs have to do with the biblical patriarchs?

. . .

There are no dates in the Bible which creates a timeline that is skewed and warped without factual data or archaeological evidence. Biblical scholars use the genealogy given in the Old Testament and work backward with ages given for dates of events. This way of dating biblical events becomes complicated the further back it goes, especially in Genesis and Exodus. The Mesha Stele from 840-850 BCE gives the dates of Mesha, King of Moab, and his history. The stele details King Mesha's interactions with Omri, King of Israel, who ruled during the 31st year of Asa's rule in Judah (2 Kings 15-16). This gives scholars a date between 870 and 890 BCE, placing the Exodus of Egypt approximately 1406-1426 BCE.

There are no chronological links between Moses and Joseph, and the Old Testament gives one line about the timeline and chronology in Exodus:

"Now there arose a new king over Egypt, which knew not Joseph." – Exodus 1:8 KJV

The details of this conundrum of accurate timelines are found in Exodus Book One. The people of Israel were too numerous, and they built Pithom and Raamses for the pharaoh. The city of Pi-Ramesses was founded during the 19th Dynasty under Ramesses II, 1292-1213 BCE. If the numbers were too great, why were there only two Hebrew midwives, Shiphrah and Puah?

Ahmed Osman helps to solve this puzzle in *Moses and Akhenaten* and notes that this short introduction in Exodus creates one timeline out of two. Genesis 46:26-27 notes when Joseph arranged for his family to settle in Goshen, outside of Egypt, there were 70 total, yet only names 69.[2] Two midwives would work for 70 people, not two cities. The stories of Moses, Exodus, and the patriarchs become mettled, flipped, and confusing when placed against archaeological evidence. In *Egyptian Origins*, Osman consistently notes that there is no archaeological evidence of major biblical characters during the reign of King David and

Solomon, creating an even more significant skew in the timeline. He remarked that Abram and Sarai would have arrived in Egypt during the 18th Dynasty, as will be explained, and that the patriarchs were intimately intertwined with the pharaohs. [3] Table 5 is provided to help understand the correlation between the pharaohs and the biblical patriarchs.

Biblical Name

Egyptian Name/ Association

Biblical Name	***Egyptian Name/ Association***
David	Thutmose III
	Amenhotep II
Jacob	Thutmose IV
Solomon	Amenhotep III
Joseph	Yuya (Father of Queen)
Moses	Akhenaten
Manasseh	Anen
Ephraim	Aye

Table 5: Table of the Pharaohs and Patriarch

THE BIBLICAL STORY AND EVIDENCE OF A GREATER STORY

One of the most essential pieces of data to point out before telling this tipsy-turvy story is that the Old Testament is written in Hebrew. The stories of the biblical patriarchs all happened before the emergence of Hebrew in the 10th century BCE out of the Phoenician script.[1] In contrast, the Egyptians wrote things as they happened, and their history gives a more accurate timeline to the Bible.

Abraham

Abraham (Abram in Hebrew) is known as the founding father of Israel to Jews, Christians, and Muslims. He traveled from Ur to Canaan with his wife Sarai, father Terah, and nephew Lot. While on their way to Canaan, they stopped in Haran for an unknown time, where god (Yahweh) met with Abram and told him that he will make a great nation (Gen 12:1-2). They continued their journey to Canaan, and after they arrived, they traveled to Egypt because of a famine. Osman notes that there was no reason for Abram and his family to travel from Ur to Canaan.[2] Yet the *Talmud* tells the story of the entire reason for Abram, Lot, Sarai, and Terah to leave Ur and journey to Canaan.[3] While chapter eleven of Genesis spends five verses explaining the history of Abram and his family, the *Talmud* places Terah as a chief officer of Nimrod, King of Babylon. Nimrod's approximate date in the biblical timeline is 2268-1868 BCE. Abram's elder brother is Charan (Biblical Haran), who was father to Lot and two daughters, Milcah and Sarai. Abraham was raised in Noah's house and lived there from the ages of 10 to 50. Abraham married Sarai, Noah's niece.[4]

Abram insulted Nimrod over the power of Yahweh versus the power of the other gods of Babylon. Due to this insult, Nimrod was furious and threw both Charan and Abram into a furnace and told them to show whose god was real, Yahweh or the Babylonian gods. Charan perished, and Abram walked on embers. Two years later, Nimrod had a dream about Abram creating a great nation. Out of fear, he ordered his guards to kill Abram. Terah learned of this plot and told Abram, Lot, and Sarai to flee to Noah's home to hide. Terah visited Abram in secret, and Abram decided they should all travel to Canaan to worship Yahweh peacefully and live safely.[5] This gives a different picture of why Abram and his family travel to Canaan.

Osman continues and notes that the journey of 700 miles was made in two parts. They first stopped in Haran, where Abram's father died, and Abram learned why he needed to travel from Ur to Canaan. This was the first time a relationship between Abram and Yahweh was shown. Abram continued with his family to Canaan. Their route was a passage for armies in the region because it was a coastal plain between Arabia and Egypt.

Canaan was not very hospitable, with an ever-drastic yearly climate consisting of scorching heat and Mediterranean winds that could reach speeds of hurricane levels. Then, rainy autumns followed by wet and bitterly cold winters, only to have a few months of gentle rain along the country's stony hills. The only agricultural product was olives.[6] Famine was common but possibly not the only reason Abram and Sarai traveled to Egypt, as the Bible says.

If Abram and Sarai had traveled to Egypt, they would have stayed in the Eastern Delta of the Nile, where wheat, barley, vegetables, fruit, and flax were abundant. Instead, Genesis notes that advisers of the pharaoh noticed Sarai's beauty and informed the pharaoh. For the court advisors to have seen Sarai, they would have to travel to the area of the pharaoh: Memphis, Heliopolis, or Thebes. Known in the Bible as On, Heliopolis was the Egyptian holy city and center of worship for the Egyptian sun god Ra, while Memphis was the trade center where the temples of Ptah and Seth were located. These two cities offered refuge to the Egyptian court to escape the summer heat. The time of Abram and Sarai's travel to Egypt would have been during the reign of the 18th Dynasty (1575-1335 BCE) when Heliopolis declined in its level of importance and Thebes became the new main center of worship for Amun-Ra.

The Bible does not give distance of travel or time spent traveling but places the characters in the location and age of people when they have children. Based on the location of the

pharaoh's residence in Thebes during Abram's time, it is safe to assume that Abram and Sarai traveled to Thebes because Genesis states:

"When the princes of Pharaoh saw her [Sarai], they praised her to Pharaoh. And the woman was taken into Pharaoh's house [in Thebes]" -Genesis 12:15 ESV

In the biblical narrative, Sarai married the pharaoh because she told the pharaoh's advisers that she was Abram's sister. When Yahweh placed disease upon the pharaoh's house, the pharaoh sent Sarai and Abram back to Canaan with generous gifts and an Egyptian servant named Hagar. They all return to Canaan safely.

"So, after Abram had lived ten years in the land of Canaan, Sarai, Abram's wife, took Hagar the Egyptian, her servant, and gave her to Abram her husband as a wife"

-Genesis 16:3 ESV

From this union, Ishmael was born. Abram was 86 at the time of Ishmael's birth. Thirteen years later, Yahweh arrived and told Abram:

"No longer shall your name be called Abram, but your name shall be Abraham, for I have made you the father of a multitude of nations" -Genesis 17:5-6 ESV

Sarai was renamed Sarah, and Yahweh told Abraham that the way man keeps his covenant with him is that every male shall be circumcised (17:10). A year later, Isaac was born.

Osman argues that Sarai became pregnant by the pharaoh and Isaac was the product of that marriage, not the son of Abram. Sarai's name was changed to Sarah before the birth of Isaac. The Biblical Hebrew meaning of Sarah is 'Princess.'[a] The name change would only make sense after Sarai's marriage to the pharaoh. If Sarah's name was changed to Princess at a later date and for the sake of the lineage that was to come from Abram, then it would make sense for Yahweh to change Abram to Abraham if it meant prince or ruler[b] but this is not the case. Thus, it must be referencing Sarah's marriage to the pharaoh, and Abram receives a dowry of sheep, oxen, male and female donkeys, male and female servants, and camels (Genesis 12:16). This custom also aligns with the Egyptian custom of sexual intercourse on the day of the wedding ceremony.[7]

If Sarah was not pregnant, why would she have a personal Egyptian servant sent with her when she and Abraham were sent away? A personal servant on top of the male and female servants that the pharaoh had already given to Abraham should have sufficed. For the father of Isaac, Osman notes that the pharaoh, not Abraham, was the father of Isaac, and this claim is confirmed in both the Talmud and the Qur'an.[10] Osman goes on to state the actual identity of this mysterious pharaoh is Thutmose III.[8] If Isaac were the son of Thutmose III, then the land of Canaan (modern-day Iraq, Syria, and Turkey) would have been an inheritance rather than a fight to claim the land in the name of Yahweh.

When Yahweh spoke with Abraham and changed his name, Yahweh promised him that his descendants would be kings and create nations (Genesis 17:6 ESV). Osman also noted that the practice of circumcision was Egyptian, dating back to at least 2686 BCE. This practice denoted

---

[a] Sarah: feminine (*sara*) denotes princess or noble lady. Masculine *(sar)* denotes chief or ruler (Abarim-publications.com).

[b] Abram: (*Ab*) father; (*rum*) to be elevated. Abraham (*abar*) to be strong or to protect; (*am*) their. Thus, 'to be their protector' (Abarim-Publications.com).

an elite class within the Old Kingdom of Egypt.[9] Josephus' writings align with Osman when he recounts the writings of Herodotus of Halicarnassus. He commented that the only people who were circumcised among mankind were the Colchians (Kingdom of Iberia, modern-day Georgia), Egyptians, and Ethiopians. Yet, between the Ethiopians and Egyptians, it is impossible to tell who began the practice.[10] A curious observation of the Genesis narrative shows the period that was placed by the narrator between Sarah in Egypt and the birth of Isaac. It even states that Sarah was barren until Yahweh appeared in the form of three people and promised her that she would give birth in a year, and she laughed.

Lastly, the sacrifice of Isaac is shown. An all-loving and all-knowing (omniscient) god, Yahweh, demands a test of Abraham's loyalty to him:

Yahweh "tests Abraham and says to him, 'Abraham...take your son, your only son Isaac, whom you love, and go to the land of Moriah, and offer him there as a burnt offering on one of the mountains of which I shall tell you'." - Genesis 22:1 ESV

Osman notes that this would not be as difficult of a decision if Isaac were not the blood son of Abraham that Genesis makes him out to be. Instead, it is a curious request if Yahweh promised to establish his covenant with Abraham through his descendants.[11] Biglino notes the oddities of this request. If Yahweh is an omniscient Elohim, why must he test Abraham? This answer rests in the fact that Yahweh does not actually know Abraham's loyalty and uses a cruel and inhumane request to prove it. Only after Abraham has Isaac bound to the altar, and Yahweh is aware that Abraham has complied with his requests, is a messenger sent to stop him.[12]

The idea of Abraham as Isaac's biological father creates the narrative of Abraham as a patriarch of Israel. But Isaac's lineage is also found in Sarah, who was a descendant of Noah through Charan. The unknown mixture of Jewish and Egyptian continues through the patriarchs.

Another version of the biblical character of Abraham is found in cuneiform writings that Zecharia Sitchin eloquently places in a story timeline. Instead of using Genesis 14:1 as the timeline to discover Abraham's life, Sitchin uses the lineage given in Genesis 13:14-29: Shelah, Eber, Peleg, Reu, Serug, Nahor, Terah, Abram, and his brothers Nahor and Harran. He places the birth of Abraham in the year 2123 BCE, ten years before Ur-Nammu became ruler of Ur and the 3rd Dynasty of Ur began. When Abraham was 75, Shulgi, the second ruler of Ur, was the catalyst of the downfall of Ur and himself.[13]

The third descendant of Shem is Shelah (*sword*), who was born 258 years before Abram, during the ascendency of Sargon in Akkad.[c] The second questionable character is Eber, from the biblical *Ibir*, which is the way Abraham identified himself (Semitic *Habiru*; Hebrew *Hapiru*; what Egyptians called the Hebrew people). Yet this translation of Habiru came from Abraham, not before. In Hebrew, the prefix '*i-*' indicates the region where someone is from. This is similar to Sumerian's prefix 'ni-'. It is common practice to drop the '*n*' when transposing the words to Akkadian or Hebrew, as is commonly seen across cuneiform transcriptions. This would then read *Nibiri,* commonly referred to as Nippur, the religious center of Sumer.[14]

Third comes *Peleg*, which means 'division' and refers to the division of Sumer and Akkad after the reign of Sargon the Great. The father of Abram, Terah, also holds a clue of Abram's origin. The Sumerian cuneiform rendering of Terah is *Tirhu,* meaning 'fate-speaker' or 'oracle-priest.' The cuneiform rendering of Abram is AB.RAM, with *Ab* meaning 'father.' AB.BA.MU meant 'governor of Nippur' when governors were the "pious shepherds" of the gods.[17] Josephus agrees with this idea and writes of Abraham as a king of Damascus.[15]

Instead of a humble shepherd, Sitchin uses the Sumerian cuneiform to discover the true roots of this patriarch. Instead of Abram having a brother named Harran, what if Harran was a location? Archaeologists

---

c *Sumerian King List* uses the Sumerian term *Agade* (meaning united) for Akkad (Sitchin. *The Wars of the Gods*, p. 246.).

found ruins of Harran located in the foothills of the Taurus Mountains of Mesopotamia. Harran served as a crossway between northwest Mesopotamia and Western Asia and was strikingly similar to the city of Ur visually, structurally, and religiously.[16] Sarai was already translated as a princess, and the name of Harran's wife, Milkha, means *queenly*. This was not the usual shepherd family as depicted in the Bible. They were a family of high status from Sumer who traveled to Canaan not to start a religion but instead called by Yahweh to gather an army and prepare for upcoming battle, as the military commander that he already was.[17]

Isaac, Jacob, and Esau

Rebekah is described as infertile, just as Sarah (Sarai) was. Osman notes that this was the way of saying that a girl was married off before she reached childbearing age.[18] The Talmud states that Sarai was the niece of Abram, who returned to his father's house to meet Sarai when he was fifty. The Talmud notes that Isaac was fifty-nine years old, and Rebekah was barren.[19] Once of child-bearing age, she gave birth to twin sons, Esau and Jacob. Shortly after the introduction of Esau and Jacob, the Bible notes that Esau gave Jacob his birthright over a bowl of stew because he "despised his birthright" (Genesis 25:34). The birthright of Esau was the inheritance he received of property or title after the death of Isaac.

The Talmud remarks that upon the death of Isaac, Esau and Jacob agreed upon Isaac's inheritance and split his land between the two sons. Jacob drew up a deed to the land from the river of Egypt to Euphrates. This deed was recorded, duly witnessed, and sealed.[20] This means that the birthright that Thutmose III gave Isaac was carried down to Jacob, his grandson. This title gave Jacob the title of Prince over the land of Egypt. Following the lineage of the 18th Dynasty, Jacob would be correlated with Thutmose IV (Table 5).

Chapter nine recalls the story of Thutmose IV, revealing that he was not the actual crowned prince or Amenhotep II's chosen successor. Thutmose IV usurped his older brother to take power, recorded this story in the Dream Stele, and justified his kingship. His justification was,

"Behold Thutmosis, my son Thothmes. I am thy father, Hor-em-akhet-Kheperi-Ra-Atum; I will give thee my Kingdom upon the earth and the head of the living. Thou shall wear the White Crown and the Red Crown upon the Throne of Geb, the Hereditary Prince. The land shall be thine, in its length and in its breadth, that which the eye of the All-Lord shines upon."[d] This Dream Stele was placed between the paws of the Sphinx after its restoration by Thutmose IV. Thutmose IV was remembered as a dreamer with a short ruling period of nine to ten years.

Thutmose IV and Jacob are both dreamers who took their right to rule within their own hands and overpowered their older brother.[21] Jacob was visited by Yahweh and renamed Israel:

"Your name is Jacob, but no longer shall your name be called Jacob, but Israel shall be your name" - Genesis 35:10 ESV

This is significant because of the way Israel breaks down: *El* and *Sar*.

- El [אֵל] – The Shining One [*Elohim*][22]
- Sar [שָׂרָה] – *sarah*, primitive root is *sar*, meaning 'to have power.'[e]

Previously noted: refers to the chief or ruler. [f]

Yahweh renames Jacob: 'Chief Elohim' or 'Elohim is Chief.' This renaming happens right before the story of Joseph is told.

The Life of Joseph: Vizier to Thutmose IV

---

[d] www.ancient-egypt.org/language/anthology/fiction/dream-stela/dream-stela---translation.html 11 Aug 2014

[e] Strong's Concordance No. 8280.

[f] Arabim-Publications.com

Joseph, like his father, was a dreamer and shared the dream of his inheritance and connection to the Egyptian royal bloodline. This is why Joseph was labeled as the favorite of Jacob's sons and received the ornamental robe. Genesis 37 remarks on the two dreams that Joseph had. He told his brothers the interpretation of his dreams—that one day he would rule over them. Joseph's brother hated him even more and sold Joseph to the Ishmaelites. He was then sold to Potiphar, the captain of the guard and officer of the pharaoh. Joseph was placed in charge of everything Potiphar owned. Potiphar's wife attempted to seduce him and he refused. Upon Potiphar's return home, she told him that Joseph attempted to rape her but she got away. Joseph was sent to prison and placed in charge of all the prisoners with such high esteem that the

"Keeper of the prison paid no attention to anything that was in Joseph's charge, because Yahweh was with him. And whatever he did, the Lord made it succeed."

-Genesis 39:23 ESV

Joseph then interpreted two dreams for a baker and the pharaoh's cupbearer. He successfully interpreted the dreams and after two years the chief cupbearer remembered Joseph and suggested the pharaoh relay his dream to Joseph for interpretation. Joseph successfully predicted a time of flourishing for Egypt along with famine and told the pharaoh how Egypt could still be successful even with the famine. After this news, the pharaoh bestowed honor to Joseph. This part of the story becomes very interesting and easily ties into Egyptian history. The timeline gets muddled because chapter one of Exodus notes that there were 400 years between Joseph's life and Moses' time. There are other ways to find the true timeline, especially since Joseph became a part of Egyptian high society:

"You shall be over my house, and all my people shall order themselves as you command. Only as regards the throne will I be greater than you.

And Pharaoh said to Joseph, 'See, I have set you over all the land of Egypt.' Then Pharaoh took his signet ring from his hand and put it on Joseph's hand, and clothed in him garments of fine linen and put a gold chain about his neck. And made him ride in his second chariot. And they called out before him, 'Bow the knee!' Thus, he set him over all the land of Egypt. Moreover, Pharaoh said to Joseph, 'I am Pharaoh, and without your consent, no one shall lift up hand or foot in all the land of Egypt. And Pharaoh called Joseph's name Zaphenath-paneah. And he gave him in marriage Asenath, the daughter of Potiphera priest of On." - Genesis 41:40-45 ESV

The signet ring that the pharaoh gave Joseph is the Seal Ring, meaning that the pharaoh gave Joseph the ring to approve everything that happened in Egypt. This is just one of the many things that the pharaoh gave Joseph, along with garments of fine linen, a gold chain, second in command, and a wife who was a daughter of the Priest of On (Heliopolis). When discovering this, Osman noted that he could not sleep when he realized Joseph's titles were the same ones that Yuya, husband of Thuya and father of Queen Tiye, held.[23] The *Talmud* gives more information about Joseph's new position within Egypt and notes that Joseph was seated upon a throne in royal clothing when his brothers arrived in Egypt.[24]

Inscriptions on the tomb of Yuya included, "the holy father of the Lord of the Two Lands [Pharaoh]," 'seal-bearer of the King of Lower Egypt," "Bearer of the ring of the King of Lower Egypt," and "Deputy of Pharaoh's Chariotry." A gold and lapis lazuli necklace with large beads was found along with the titles. Grafton Eliot Smith also examined Yuya's mummy in 1905 and noted that his cranium was not Egyptian, and the structure of his nose was more common in Europe than in Egypt. Even more fascinating were the various forms his name was written, all denoting the Hebrew letter "Yod," short for Yahweh. Osman even lists how it was written: "Ya, Yi-Ya, Yu-Ya, Ya-Yi, Yu, Ya-Y, Yi-Ya, Yu-Y". In the 3rd century BCE, the Egyptian historian Manetho wrote about Yuya as "*Sef.*" The pharaoh gave Joseph an Egyptian name starting with Zef, which could be a translation from "*Sef*," and the

Hebrew name for Joseph is *Yosef,* compounding the Hebrew *Yo* and Egyptian *Sef.*[25]

The author of Genesis continues to play tricks on the reader because, in chapter 45, there is a snippet hidden in the story from Joseph that makes this connection obviously clear. Joseph sends his brothers back to Israel (Jacob) with a message to tell him. Within this message, he says:

"So it was not you who sent me here, but Elohim.[g] *He has made me a father to Pharaoh*, and lord of all his house and ruler over all the land of Egypt." -Genesis 45:8 ESV, emphasis added.

Yuya was the vizier to Thutmose IV who married Thuya. His origins are unknown but are commonly believed to be from other regions outside of Egypt. Yuya and Thuya had a daughter, Tiye, and two sons, Anen and Aye. Tiye grew up alongside Amenhotep III and became the Great Royal Wife. Joseph had two sons, Manasseh and Ephraim, who received odd blessings from Israel before his death. Genesis 48:20 notes that, "he [Israel] put Ephraim before Manasseh." Joseph tried to stop, but Israel just said that he was about to die and there was nothing that he could do about it because it was already done. This story is weird when read without context. Still, Egyptian history shows that Anen (Manasseh) became a chancellor of Lower Egypt, a second prophet of Amun, and a priest of Heliopolis. In contrast, Aye (Ephraim) became a pharaoh after the death of Tutankhamun. As the next chapter will explain, all the Amarna kings descended from the line of Joseph.

While reading this chapter, it is probably impossible for questions not to arise, but before moving to the next chapter, which will produce more and more questions, let us summarize. First of all, this is only one viewpoint. I use multiple sources, including literary ones such as the Qur'an, Talmud, and Manetho, and Egyptian archaeology. Dr. Kenneth Kitchen, Professor of Egyptology at Liverpool University, does not

---

[g] Name obtained from *Names of God Bible* version.

support this viewpoint. He regards the Bible as an accurate historical account and accepts the Hebrew Bible's accuracy over hieroglyphics, believing ancient Egyptians exaggerated their statements. Osman points out that to become a Professor of Egyptology, Dr. Kitchen obtained his PhD in Bible studies.[26] To discover the truth, one must look beyond bias even when the possibility of truth becomes uncomfortable.

Moses: Biblical Patriarch or Egyptian Pharaoh?

Establishing the Narrative in Exodus

Exodus begins with naming the sons of Israel (Jacob), who moved to Egypt with Joseph (Exodus 1:1-6). This is a reprieve of Genesis 46:5-26, which states there were 70 people in all that moved to Goshen, outside of Egypt, but still only gives 69 names. Osman notes his opinion and thinks that the 70th person was Tiye because she was already in Egypt when the Israelites arrived at Goshen.[1] While it is only speculation, the argument does make sense. The people grew in numbers, and a new pharaoh arrived who did not know who Joseph was (Exodus 1:7-8). This particular statement points to two different points: the time of entry into Egypt and a new pharaoh who did not know about Joseph.

Exodus 12:40 ESV notes the time between the arrival of Jacob and the new pharaoh was 430 years. This same verse in the *Septuagint* mentions that the time spent in Egypt and the land of Chanaan (Canaan) was 430 years, moving towards a narrower timeline. Josephus gives more clarity to this: "They left Egypt in the month of Xanthicus, on the fifteenth day of the lunar month: four hundred and thirty years after our forefather Abraham came into Canaan, but two hundred and fifteen years only after Jacob removed into Egypt."[2] The *Talmud* backs up Josephus and notes that the Hebrews lived in Egypt for 210 years.[3] With the assistance of the *Talmud* and Josephus, the two different points of time are approximately 120 years apart.

The *Talmud* gave a history of this 210-year period when this new pharaoh came into power, 102 years after Jacob entered Egypt. Jacob lived in Egypt for 17 years before his death at the age of 147.[4] Fifteen years later, the Pharaoh Joseph served, Thutmose IV, died, and 71 years after Jacob entered Egypt, Joseph died at the age of 110. Then, Pharaoh Amenhotep III ruled Egypt and governed wisely. In the next 15 years, the rest of Jacob's brothers died, with Levi being the last, 86 years after Jacob entered Egypt. Osman notes that Joseph, also known as Yuya, was a vizier to Thutmose IV and Amenhotep III, as the *Talmud* verifies and reports that "a new pharaoh ruled wisely after Joseph died."[5]

This decreases the timeline to approximately 124 years. The conjectural dates of the 18th and 19th Dynasties show that Amenhotep III's and Ramesses I's reigns were approximately 97 years. This is an approximation because no exact time is given due to the variations of dates given for reigns in Egypt. Yet now there is a 27 year gap that provides more comfort than 430. But what else can show the variation the author uses in their timelines?

Exodus helps the reader understand when Moses lived because it notes that the Israelites built Pithom and Raamses. This text could refer to two cities. Pithom, also known as the "House of Atum," was built during the 18th Dynasty and was a short-lived settlement. The other possibility is that Pithom and Raamses split Pi-Ramesses, the capital built during the 19th Dynasty by Ramesses II.[6] In the *Talmud,* the pharaoh also states that these cities were refortified during his reign but that it was a call to all peoples of Egypt, and the workers were given a livable wage.[7] The hard evidence of Pi-Ramesses being built or refortified during the reign of Ramesses II provides evidence that the Exodus took place during his reign. This, however, does not explain the initial problem of two timelines within one story; what else can help prove that this story is vague for a purpose?

The jarring evidence is when the pharaoh speaks to two Hebrew midwives, Shiphrah and Push.[8] Osman notes that the pharaoh would have spoken to the two women in person and, therefore, would have lived near Goshen.[9] The pharaoh tells the two midwives to kill the males and let the females live. This does not work; the pharaoh gets angry, and the people multiply and grow very strong. The pharaoh orders a degree to cast all male children into the Nile.[10] The *Talmud* remarks it was a decree sent out among the land and not spoken to two midwives.[11] If only two midwives were present, then this would refer to the time that the entirety of the Hebrew civilization in Goshen only needed two midwives. If a population grew to the size Exodus notes as "very strong," it would require more midwives than two. This point helps to solidify that the vague story the writer is attempting to tell is of two different timelines.

The real question becomes, why is there a need to vaguely tell the story and confuse the reader? Osman notes that this was done deliberately to hide the fact that Moses was born an Egyptian, and the changes at the story's beginning are created to hide this. Chapter two of Exodus notes that Moses was born to a humble Levite family, which the reader is given the name of in chapter six.[12] This is the first indication that something is off because the Bible is very particular about genealogies and places them before the birth of the character to be spoken about. It is also interesting to note that the name of Moses' father is Amram. The reign of Akhenaten and the following four pharaohs are known to history as the Amarna kings.

Why the Amarna Kings are not Spoken About Much

Before jumping straight into the deep end and recalling information from chapter ten, the destruction of Egyptian custom by the Amarna kings was great. This caused the names of the Amarna kings to be intentionally omitted from the *Turin King List* at Abydos. Osman consistently argues in his books that this is one—if not the—reason that looking back into history we see two separate characters merge into one: Akhenaten and Moses. The stories of Akhenaten, his new Aten religion, and his loyal following were to be silenced. This silencing caused many to tell similar stories with different names; over time, one name emerged over all the rest.

Josephus recounts the history of Manetho, an Egyptian priest and historian from the 3rd century BCE. He notes that Manetho gave a fictitious name to King Amenophis, who had incredible stories. No accurate data for this king was written about, which was not normal for Manetho. The people of Avaris were shepherds, and they appointed themselves a ruler from the priests of Heliopolis named Osarsiph. The shepherds all pledged an oath of obedience to him. Osarsiph ordered that they should worship no Egyptian gods, along with setting laws in place that opposed Egyptian custom. Osarsiph prepared them for war with Amenophis. The people from Avaris attacked the Egyptians, overthrew the cities, burned the temples, and acted like complete barbarians. This priest of Heliopolis was Egyptian by birth and was named Osarsiph (from Osiris) and changed his name to Moses.[13]

The Birth and Early Life of Moses in Egypt

The only information Exodus gives about Moses' birth and childhood is that the servant girl of the daughter of the pharaoh rescued Moses. Moses was then taken to the Hebrews to be nursed, and when the child grew up, he was taken back to the daughter of the pharaoh, and she gave him the name Moses.[14] The *Talmud* notes Moses was the youngest of his siblings, Mir'yam and Aaron.[15] Moses is given many names in the *Talmud*. Bithia, the pharaoh's daughter, names him Moses, and it notes that his mother gives him a different name, Yekuthiel.

- Yekuth – possibly derived from [יקה] *yaqa*
    - (verb) to be pious or care for religious duties
    - (noun) obedience
- El – [אֵל] *god*[a], *Elohim*

---

[a] Strong's Concordance No. 410

- Yechavel [יְחִיאֵל] – pronounced (yekh-ee-ale) means *may god live.*[b]

Similarly, A*kh*en*aten* (emphasis added) means *creative manifestation of the Aten,*[16] the Egyptian god that Akhenaten brought forth in his reign. Osman also notes that *mos* is Egyptian for child,[17] and using the name Moses is a reuse of the Egyptian word for son.

While Exodus states that Moses was born and grew up with the Hebrews, the Talmud gives more information about Moses' childhood. It tells the story of when he was three years old, he was seated at the table with the pharaoh, his queen, Bithia (his mother), and his two uncles. The pharaoh placed Moses on his lap. Moses then reached up and grabbed the crown from the head of the pharaoh and placed it on his own head.[18] This angered the pharaoh, and he sought advice from his royal advisors Reu'el, Job, and Bi'lam.

The story continues in the *Talmud* and aligns with Exodus as the pharaoh was angry about the increasing numbers of Hebrews in the area and asked his advisors, Reu'el, Bi'lam, and Job, for advice. The pharaoh became angry with Reu'el and Bi'lam's advice and sent them away. Bi'lam joined Kikanus of Ethiopia, and Reu'el took Joseph's staff to Midian.[19]

There was a rebellion and King Kikanus of Ethiopia appointed Bi'lam his representative. At the age of 18, Moses killed the Egyptian, fled to Ethiopia, and joined the army. He became the king's favorite and, after his death, was appointed the king and leader of Ethiopia. He marries the widow of Kikanus, Adonith, and rules as king over Ethiopia, but Adonith wished her son of Kikanus to rule. Moses stepped down as king and traveled to Midian, where he met Reu'el and married Zipporah.[20]

---

[b] Strong's Concordance No. 3171

Akhenaten and Moses: Two Characters Collide

This is where the parallels between Akhenaten and Moses begin to emerge. Akhenaten is known as the renegade pharaoh because he brought about the worship of the Aten and attempted to transform the Egyptian worship of many gods into the worship of one god, Aten. The *Talmud* notes that Moses went to Ethiopia and was given the crown and the king's widow, whose name was Adonith, which, translitering from Hebrew to Egyptian, becomes *Aten-it*.[21] The Hebrew *d* and *e* are transformed from the Egyptian *t* and *o*. This is also seen in the Jewish creed *"Schema Yisrael Adonai Elohenu Adonai Echod,"* meaning "Hear, O Israel, the Lord thy God is one God." This transliterates to Egyptian as "Hear, O Israel, our God Aten is the only God."[22]

Exodus aligns with the *Talmud* and continues with the story of Moses slaying an Egyptian who was beating a Hebrew. He killed the Egyptian and then hid him in the sand. For fear of his life, he fled to Midian. Moses met up with the priest of Midian, who had seven daughters. Moses stopped the shepherds attempting to drive the daughters away. The daughters took him to their household, and he met their father Reu'el. Moses married Zipporah, Reu'el's daughter, and she gave birth to Gershom (Exodus 2:11-22). Exodus 3:1 changed Reu'el to Jethro, the priest of Midian. While Moses was in Midian and Ethiopia, the pharaoh died and a new pharaoh, the second son, ascended the throne.

Moses then encountered the angel of the Lord in a fiery bush, but the bush was not consumed. This fantastical burning bush represents the holy spirit or visual manifestations of god.[c] There is something more concrete, however, as noted by Biglino. He remarks that the Hebrew word used for bush is '*seneh*' and also means 'rocky ridge,' with the root of the word having to do with sharpness. He notes that this sharpness can be associated with a sharp-edged rock or geological formation, or even a mount.[23] It is easier to believe that a rock's surface could be aflame but not burn.

---

c Biblestudytools.com, scripturalthinking.com

The Lord then tells Moses on this rock that he will rescue the people of Israel:

"But Moses said to Elohim, 'Who am I that I should go to Pharaoh and bring the children of Israel out of Egypt'?" – Exodus 3:11 ESV

"God said to Moses, 'I AM WHO I AM.' And he said, 'Say this to the people of Israel: "I AM has sent me to you'." – Exodus 3:13 ESV

The Jewish Torah translates 'I AM WHO I AM' as '*Ehyeh asher Ehyeh.*' The verse notes that *Ehyeh* is the singular form of Yahweh. While *asher* is normally translated as who, Sumerian *ash* means *perfect* or *one*. This phrase then translates to 'I am the perfect one' or 'I am the first [of the Elohim].'[24] Moses protested because the people would never believe him. In Exodus 4:2, Yahweh responds to his protest and asks, "'What is that in your hand?' Moses said, 'a staff'." The *Talmud* notes the staff Moses held was the one Adam carried out of Eden and Noah inherited, then passed through Abram, Isaac, Jacob, and Joseph.[25]

This same staff is used when Moses and Aaron speak to the pharaoh. Aaron throwed down the staff that Moses held and it turned into a serpent. The Pharaoh summoned his wise men and sorcerers, and had each cast down their staffs to see if they would become serpents. Aaron's serpent staff swallowed up all the wise men and sorcerer's staffs.[26] Osman notes in the Qur'an Sura VII, 104-124 that the staff represents a symbol of authority instead of a magician's trick. He also notes that Egyptian kings had a collection of rods that represented their authority. The Hebrew word used to describe Moses' rod was *nahash,* which means *serpent* or *soothsayer*—euphemistically, he who solves secrets. The Hebrew word for copper is *nehoshet,* and Moses' staff was called *nahash nehoshet*, or a copper serpent.

. . .

"He removed the high places and broke the pillars and cut down the Asherah. And he broke in pieces the bronze serpent that Moses had made, for until those days the people of Israel had made offerings to it (it was called Nehushtan)." – 2 Kings 18:4 ESV

This was the staff that was kept in the Temple of Yahweh until, as the passage reveals, it was destroyed by King Hezekiah.[27] This staff is depicted in the tomb of Kheruef on the relief of Queen Tiye. In the relief, the fourth official stands in line to hand items to Amenhotep III to prepare for the Sed celebrations; this official holds the curved scepter with a serpent's head in his left hand.[28]

The Magic of Moses

The magical acts that Moses performed were ritual rites performed by the king to establish his authority. Osman notes that when the Nile turns red, it indicates a specific time of year, the Inundation; near the end of summer, the Nile turns a red hue. The yearly Sed festival celebrated this season. There are satellite images today that show this phenomenon. The yearly Inundation of the Nile allowed life to flourish just as the light of Aten worked through Akhenaten to help his people flourish.[29]

It is interesting to note that while the Bible speaks of all types of miracles and things that seem magical, this belief in magic is prohibited by the Israelites, just as it is forbidden to make any idol (which will be explained). This is because, as Osman notes, magic implies a separate realm from that of the deities and nature. Magic is an attempt to use occult or hidden forces that do not rely entirely on a deity. This is, therefore, in contrast with Biblical and Egyptian monotheism.[30]

This confrontation of Moses shows that pharaoh is not just letting the Israelites out of Egypt, but Akhenaten returned to Egypt from Midian to state his rightful place on the throne. If Akhenaten were forced to abdicate the throne, as history shows after he attempted to create a monotheistic religion of Atenism, he would have tried to return and claim his rightful place to rule after the death of Horemheb. He would have had his royal scepter with him when he left and returned with it to Egypt when he attempted to state his rightful place on the throne instead of Pa-Ramses (Ramesses I). The wise men understood his rightful rule, but Ramesses I refused it and sent his army to force him and his followers out of Egypt.[31]

The parting of the Red Sea in Exodus 13:18 is a mistranslation that leads to another magical act of Moses and Yahweh. 'Sea' in *Brown-Drivers-Briggs Hebrew and English Lexicon* translates as *sea*,[d] while the Hebrew word *suph* is commonly mistranslated as 'red.' The proper translation for this context would be *reeds.*[e] The sea of reeds is at the upper end of the Gulf of Suez and is shallow and marshy.[32] As noted in Exodus 14:21, a strong east wind would have made this shallow marsh dry enough to walk over but not dry enough for the pharaoh's chariots. Geologists note that, during this time in history, there was an increase in tectonic plate movement that caused a tsunami in the region, causing the natural water dispersion to be altered. This massive decrease in water height is also seen after storm surges, similar to Hurricane Irma in 2017.

The plagues are not the only time that the magical acts of Yahweh are shown in Exodus or throughout the Old Testament. As with the parting of the Sea of Reeds and the Nile turning blood red, these incredible acts can be described using modern knowledge and terminology, as well as UFO and extraterrestrial contact.

ANOTHER VERSION OF MOSES

It would be wrong to give only one other opinion on the character of Moses. As Osman pointed out, it is impossible to truly know who Moses was because it happened so long ago. George Smith notes similarities between names found in Babylon during the reign of Hammurabi that correlate with those of the Bible, Abuha son of Ishmael, a witness to legal documents. Smith also noticed a correlation between Moses and

---

[d] Strong's Concordance no. 3220

[e] Strong's Concordance no. 5488

Sargon I, the King of Akkad.[f] The "Kouyunjik Tablet" reads much like the Exodus story of Moses' birth. Sargon says that his mother was a princess, he did not know his father, and his uncle was the ruler. After he was born, his mother placed him in an ark made of rushes and bitumen and put him in the river. The river took him to Akkad, where he became king.[33]

It is impossible to truly reveal the origins of the biblical patriarchs. Only history reveals the truth, and when history has been rewritten, the truth is lost. Whether the story of Moses is allegorical and taken from the story of Sargon I or hides the correlation to Akhenaten is unclear. One point of interest is the fact that the entire Pentateuch was revised. Ezra took it upon himself to take all the stories from antiquity, pull out various characteristics from kings and other important figures, and then create a fictional story that supported the agenda of the Temple Cult that became known as the Jews.[34] Yet the stories that Ezra created were older versions of stories. Moses, like Akhenaten, taught one god and one truth at the exclusion of everything else. The stories that spoke of Moses were stories taken from the cultures around them: Egypt, Canaan, Philistia, Jericho, and Asia.[35] This caused waters where the actual truth lies to become more mirky. More information is then needed to discover the real story being told.

---

[f] Smith places the reign of Sargon I around 1600 BCE. Blavatsky references Professor Sayce, who argues the reign is much older and Sargon I could have reigned at least 2000 years before the time of Moses (Blavatsky. *The Secret Doctrine Vol I*, p. 240.).

Mysterious Aspects of Yahweh with Moses

Magical things concerning Yahweh are often translated as the "glory of god." This is a mistranslation that has significant ramifications. The Hebrew word for glory is *kavod,* which is a mistranslation. *Kavod* is related to *kaved*, meaning 'heavy,' and is commonly associated with the weight or heaviness of glory. When glory is used in specific areas of the Old Testament, very concrete details go along with descriptions, causing the definition of *kavod* to be something other than glory.

The second word that causes mistranslation and confusion is spirit, *ruach*. This mistranslation has been used since the 3rd century BCE with the first Greek translation of the Septuagint. Before this translation, the Alexandrian authors translated *ruach* as *pneuma*. *Pneuma* is the Greek word for 'breath,' 'wind,' and *pneumo* (breath of life), with the last rendering being used to transform the *ruach* into *spirit*. The Hebrew use of *ruach* gives a different view with physical properties: *breath, wind, storm wind, rushing wind,* and *breath of air.*[a] This also provides a new

[a] Strong's Concordance no. 7307.

outlook on how Yahweh is portrayed when his presence or the presence of angels is in a cloud.

Clouds of Yahweh

As the Israelites walk through the Sea of Reeds, an angel of God is before them, and Yahweh is in a pillar of cloud behind them. During the day, there was a pillar of smoke, and at night, a pillar of fire. Madden Jones notes that this cloud associated with Yahweh is always commented on as Yahweh being in the cloud and not that Yahweh is associated as the cloud or mass of clouds.[b] This use of the cloud is an aspect of the opaqueness and the ability of the cloud to float, glide, hover, and move. The pillar is associated with the shape of clouds or fire, as a type of conical formation or, more accurately, pillar.[1]

This is not the only time that Yahweh is associated with a pillar of smoke, a pillar of clouds, or a pillar of fire. Exodus, Numbers, and Deuteronomy all reference this cloud that Yahweh is in.[c] This is why the *glory* or *spirit of God* has such thunderous and majestic attributes. These attributes could then be reassociated with this new idea that Yahweh hovers or glides in something that produces fire and smoke, or the cloud could be a term used for a flying machine[2] or maybe even the UFO[d] that Yahweh travels in.

O'Brien comments on this and notes that the cloud that Yahweh travels in has a solid exterior and illuminated interior. The pillar of fire and smoke is described by a group of people who have never seen this type of illumination before. From this location of Yahweh, he could observe all that was happening around the Israelites.[3]

---

[b] Strong's Concordance no. 6051.

[c] Exodus 13:21, 14:19-20 & 24, 33:9-11, 40:38, Numbers 12:5, Deuteronomy 31:15, 33:20-23

[d] Genesis 1:2 explains the ruach of the Elohim hovered over the faces of the waters, *ruach* translates to RIV which translates to UFO.

MOUNT SINAI LANDING

The next instance of Moses and Yahweh is on Mount Sinai. This is seen as a fantastical physical manifestation of Yahweh, where trumpets are heard, and smoke and lightning are seen. These incredible descriptions are often attributed to the *kavod*. Still, there is something very physical about actual smoke, fire, the shaking of a mount, lightning, and the threat of being killed just from touching the mountain. If this language were used in today's society, a standard reference would be something similar to a rocket that is landing with radiation that kills those close to it.

This is precisely what is described in Exodus. Yahweh informed Moses that he, in his craft, would be landing on Mount Sinai to speak with him. He then gives instructions for the people before this landing: (1) wash garments (2) abstain from sexual relations (3) must be consecrated for two days (4) set markers around the mountain (5) wait for the trumpet blast (6) do not touch the mountain (Exodus 19:10-20).

The landing or descent of Yahweh is described as thunders and lightning, loud trumpet blasts, and fire descending while smoke covers the entire mountain and then travels upward like a kiln, causing the mountain to "tremble greatly" (16-19). With the knowledge of rocketry and propulsion, this language is easily understood today when the mystical veil of the *Bible* is taken away. Descending fire and copious amounts of smoke become the firing of retro rockets. The loud noises become roaring engines, and the lightning becomes electrostatic discharges.[4] The Greek translators of the Old Testament were aware of the electrostatic properties of the electron, and the visual description of lightning and electrostatic discharges would have been understood in antiquity.[5]

Once this fiery scene has concluded, Moses receives the Ten Commandments and speaks with Yahweh:

"Moses, Aaron, Nadab, Abihu, and seventy elders of Israel went up [the mount] and saw the God of Israel. There was under his feet as it were a pavement of sapphire stone, like the very heaven for clearness." - Exodus 24: 9-10 ESV

The feet described by the author could be the landing gears seen today on the self-landing rockets that first touch the ground.[6] The Hebrew origin of foot (*regel*) is of unknown derivation.[e] *Regel* is translated into Ethiopic as 'vehicle.' As for the pavement of sapphire stone, Madden Jones remarks that the craft could have had a multi-colored lighting effect that reflected off the rock the craft landed on or could have been the color of the craft's underbelly.[7]

For the priests to come up the mountain, Moses informed them of the instructions from Yahweh to cleanse themselves. This type of "sanctification" process could have been an ancient use of various substances to place on the skin with a particular type of binding liquid to protect the priests and Aaron from the radiation of the ruach of Yahweh. Madden Jones speculates on the kind of radiation released from the ruach. The type of propulsion needed for an interplanetary craft is nuclear energy. While the energy used by the Elohim is unknown, Exodus details the effects that the people experienced from the unknown type of radiation. Some possible sources of radiation are microwave or electromagnetic radiation. The people place oils on their skin, abstain from sexual relations, wash their clothes before and after contact with this radiation, and are advised not to go near it because of certain death.[8]

Moses spends forty days and forty nights with Yahweh and then receives the two tablets. When he travels down Mount Sinai in Exodus 34:35, he covers his face with a veil because of "the shining of his skin." The "shining" of Moses' skin is not because of holiness but rather an effect of ionizing radiation.[9] These descriptions are just the beginning of radiation and biological hazards associated with Yahweh and his *ruach*. After Moses had spoken with Yahweh for forty days, he began to build the Ark of the Covenant and the tabernacle.

---

[e] Strong's Concordance no. 7272.

The Ark of the Covenant and Tabernacle: An Advanced Communication Station

Exodus 36-40 details the tabernacle, the Ark of the Covenant, the lampstand, priestly garments, and incense. The general picture that emerges is a gold-covered ark seated inside a structure 15 feet high, 45 feet long, and 15 feet wide, covered in gold. Everything within the holy of holies and courtyard of the tabernacle is covered in gold, and this all sits atop a foundation of 9,600 pounds of silver.[10] This is all placed under a tent (the tabernacle) that is 75 feet by 150 feet. Between each of the curtains was a brass-plated pillar anchored into the ground. There are golden cherubim interwoven in the linen of the tabernacle. When the spirit of Yahweh hovered over the tabernacle, Yahweh communicated with the people:

"Now on the day that the tabernacle was raised up, the cloud covered the tabernacle, the tent of the Testimony; from evening until morning, it was above the tabernacle like the appearance of fire. So it was always: the cloud covered it by day, and the appearance of fire by night. Whenever the cloud was taken up from above the tabernacle, after that the children of Israel would journey; and in the place where the cloud settled, there the children of Israel would pitch their tents."

– Numbers 9:15-17 NJKV

The description of the cloud, as has been shown, was the *ruach* of Yahweh. These verses describe a hovering craft's movements as it moves around the wilderness. This same description is found in Egyptian descriptions of fiery circles that appeared over his palace. Mirroring the Eye of Horus ascending.[f,11]

Osman makes an almost glass-shattering revelation about the cherubim

---

f Description of the ascending Eye of Horus or Eye of Ra is found in *Ancient Records of Egypt Vol II*, p. 131.

atop the Ark of the Covenant and recalls the Ten Commandments from Exodus 20:4: "You shall not make for yourself a carved image or any likeness of anything that is in heaven above, or that is in the earth beneath." This is important because, as Osman notes, the word *cherubim* occurs 91 times in the Hebrew Bible, yet a description of the *cherubim* is never given.[12] Today, these cherubim are regarded as angelic beings or guardians with wings. They have human, animal, and bird-like qualities, as noted in Ezekiel (more to be explained about Ezekiel in chapter nineteen). Why, if Yahweh explicitly states as the second commandment that the Israelites are not allowed to make a carved image of anything in heaven or on earth, does he instruct Moses to make two cherubim that will sit atop the ark?

Exodus showed that Yahweh gave Moses very specific instructions. He was even told multiple times to ensure that he followed the instructions.[g] Mauro Biglino comments that this precision is not just because it is the Ark of Holiness as it has come to be seen throughout the ages. Instead, the specificity of it points to something completely different. He even mentions that Yahweh showed Moses a precise model of the Ark, and he was to follow the blueprints of this model.[13] Yahweh indicates that the Ark of the Covenant, the mercy seat, and the cherubim had specific designs, positioning, and functions.[14] Yahweh tells Moses to build this Ark and tells him its purpose will be revealed. In Deuteronomy 10:15, Yahweh reveals that it will become a place where the Tablets of the Law are held.

Moses was instructed to make the Ark out of acacia wood, to be 2.5 cubits long, 1.5 cubits wide, and 1.5 cubits high (approximately 45 inches long by 27 inches wide and 27 inches high).[h] He is told to cover the inside and outside in gold. Beneath, he was to construct four gold rings as the feet of the box, two on the left and two on the right. Moses was to get two rods made of acacia wood and cover them in gold, place them in the rings, and never remove them.[i]

---

g Exodus 25:9; 25:40

h One cubit is approximately 18 inches.

i Exodus 25:10-15

Moses was then instructed to place the mercy seat on top of the Ark as a foundation for the cherubim. This mercy seat had the exact dimensions of an ark, 2.5 cubits long by 1.5 cubits wide, and was to be made of pure gold. The Hebrew word for 'mercy seat' is *kaporet* which means to cover and protect.[15] The Church may interpret 'mercy seat' as a reference for covering sins. Instead, take off the magical scripture goggles, and it is easy to see a solid gold lid that perfectly fits onto the ark, protecting the Tablets of the Law that were to be placed inside. Atop the lid sit the two cherubim.

The cherubim, or *kerubim* as they are translated in Hebrew, are fascinating. The only verse that describes the cherubim wings is verse 18, which describes them as beaten work. The origins of the phrase 'beaten work' show that this gold was to be molded and hammered, knotted into a coil-like formation over the top of the lid.[16] While modern-day renderings show two angelic beings or cherubim outstretched on top of this mercy seat, no distinct placement is mentioned. It just notes that they are supposed to face one another, and the wings are supposed to be outstretched over the mercy seat.

While this may seem confusing, a look into the origin of 'wing' can create clarity. Wing translated back to Hebrew is *kanaf,* which has multiple meanings: 'cover and conceal from view,' 'cover,' 'protect,' and 'wing.'[17] The purpose of the mercy seat is plainly stated by Yahweh:

"There I will meet with you, and from above the mercy seat, from between the two cherubim that are on the ark of the testimony, I will speak with you about all that I will give you in commandment for the people of Israel." – Exodus 25:22 ESV

This is not some etheric meaning either. Yahweh is being very literal and will speak with Moses.

The location of the Ark of the Covenant today has become a legend. One story notes that the Ark was taken to the island of Elephantine around the 6th century BCE, and then, in the 3rd century, it was brought to Aksum and kept in the Church of Our Lady of Zion. The story states that it is still there. Biglino mentions the story of Italian architect Professor Giuseppe Claudio Infranca and his travels to Aksum. He was invited by the clergy to visit the Saint Mary of Zion shrine and entered the Holy of Holies, where he took a photo of the ark. As he takes the photo, he has an unknown ringing in his ears. As reproduced here, Biglino and his team show the cherubim atop the Ark in the photograph. This depiction creates a different perception of the Ark and the so-called wings to present a new depiction of panels or electrodes rather than angelic wings.[18]

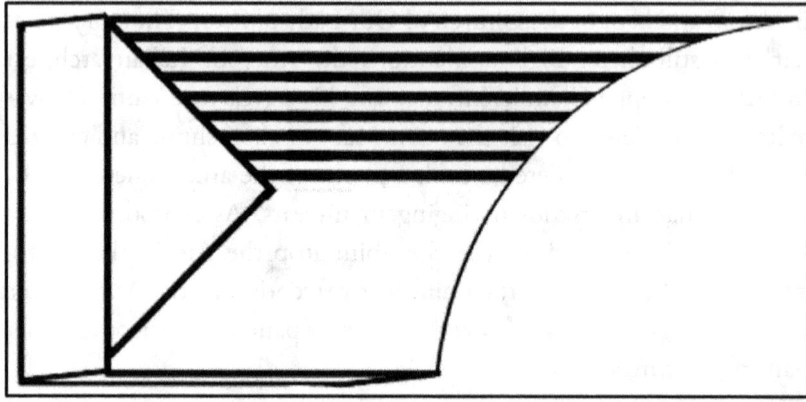

Figure 18: Cherubim

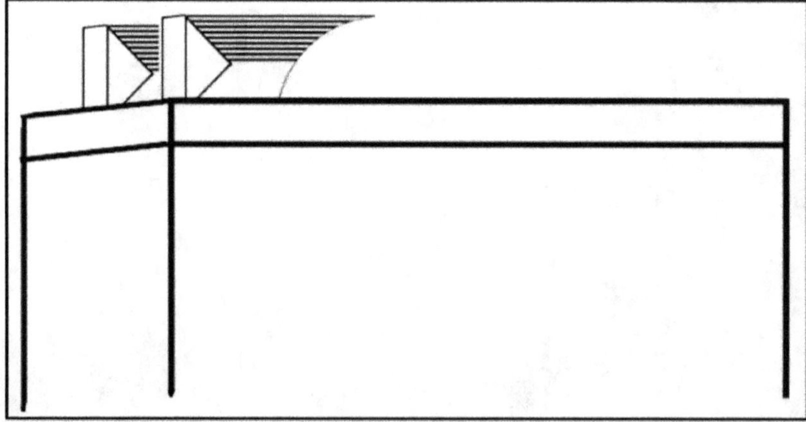

Figure 19: Ark with Cherubim Positioning

This means that the mercy seat and cherubim are a type of transmitter that allows communication between Yahweh and Moses. Rabbi Moshe Levine perfectly explains the specific construction of the Ark and its use of acacia wood in *Le Tabernacle,* as Mauro remarks. He explains that the acacia wood acts as an insulator while the gold covering both the inside and out acts as the two containers, creating an electrical capacitor.[19]

Not only was this Ark used as a verbal telecommunications device, but Leviticus 16:2 shows another feature of the ark. Yahweh tells Moses that he "will appear in the cloud over the mercy seat." The use of the word "appear" reveals that Yahweh is using the Ark as a type of holographic communication device that is seen in science fiction films today.[20] The Christian angelic depictions of the Ark and cherubim have nothing to do with the incredibly advanced technology spoken about in Exodus.

The Dangers of the Ark and Tablets

The Ark was not to be messed with. Almost 65,000 deaths were associated with the misuse of the Ark and tabernacle.[21] The *Book of Samuel* chronicles one instance of a bystander touching the ark:

"And when they came to the threshing floor of Nacon, Uzzah put out his hand to the ark of God and took hold of it, for the oxen had stumbled. And the anger of the Lord was kindled against Uzzah, and God struck him down there because of his error, and he died there beside the ark of God." – 2 Samuel 6:6-7 ESV

These verses alone present a dichotomy of both Yahweh and the ark. If Yahweh is an all-loving god as portrayed in the Bible, then why, if Uzzah is trying to protect the Ark from falling, would Yahweh kill him? However, if the Ark has nothing to do with Yahweh and is, in fact, a high-voltage electrical capacitor that could electrocute anyone who touches it, then this passage makes sense. It also makes sense why Joshua tells the Israelites to stay 2,000 cubits (about 0.6 miles) away, and as noted earlier, why Joshua told the Levite priests to consecrate themselves before taking up the Ark to cross the Jordan River.

Along with the dangers of electrocution, many of the deaths attributed to the Ark and the tabernacle deaths due to radiation poisoning are seen in Exodus 32. Mild radiation poisoning is seen when Moses comes down from Mount Sinai after speaking with Yahweh, and his skin shines. This is the first instance of the radiation associated with Yahweh and the tablets. The next instance is after Moses discovers the golden calf and angrily breaks the tablets of testimony. After he orders the death of 3000 people, a plague breaks out and kills many of the Israelites, and they are commanded to leave Sinai.

When the Philistines attempted to take possession of the Ark in 2 Samuel 5, the people within the entire city of Ashdod developed tumors. They tried to take the Ark to Gath, then to Ekron, and finally sent it back to the Israelites. Fifty thousand people died in three cities, and the rest were left with tumors and lesions. Other examples of death due to radiation include the fire that killed Aaron's sons Nadab and Abihu in the tabernacle after using a different type of incense (Leviticus 10:1-3); the earth-shaking explosion that killed 250 men and Korah; and the strange plague that resulted after a fire broke out and the censers "became holy," killing 14,700 people from a mysterious cloud that came from the Tent of Meeting (Numbers 16).

Yahweh tells the Levite priests to consecrate themselves for seven days before beginning their work in the tabernacle. They obey and spend seven days at the door of the tabernacle to "consecrate" themselves. As the Ark was emitting radiation, the priests would have slowly adjusted to the radiation over the weeklong period. This "consecration" served the priests when they interacted with the Ark face-to-face, and their bodies would build a tolerance against its harmful effects.[22] Various oils and spices are used to create the anointing oil, and this technique is used today to develop acoustic holography is similar.[23]

Further examples of deaths caused by radiation emitted from the Ark include the use of a scapegoat to test radiation levels after the death of Aaron's two sons; the death of some men in Beit Shemesh after looking upon the Ark from 1 Samuel 6:19; and the leprosy that Miriam and King Uzziah developed after interacting with the Ark without the proper procedures found in both Numbers 12:10 and 2 Chronicles 26:19-21.

Ann Madden Jones argues the radiation or clouds connected with the deaths were clouds created by radon gas. Radon is the product of

Radium decay and is colorless, tasteless, and odorless. Concentrated radon gas has high levels of radiation. The real question is whether the tablets and the Ark produced a mysterious virus that killed many people over hundreds of years or whether the Ark emitted radioactive particles and radon gas.[24]

Today's scientific advancements give readers better insight into the accuracy of handling radioactive elements. Locations that house radioactive materials have strict instructions for the few people who work with radioactive materials. These instructions include handling materials, donning and doffing gear, and decontamination. The small group of Levites, such as the temple priests, is the small group of workers used today.[25] Events such as Chernobyl are examples of what happens when mistakes are made with radioactive materials. When mistakes were made in the Old Testament, more than 60,000 people died. This description makes more sense than a god who smites people because of his "holiness."

The entire argument over His actions being holy or unholy becomes obscured after the word "holiness" is understood to refer to a radioactive element. Haran gives more information regarding holiness and contagious holiness. He notes in Exodus 29:37 and 30:29 that the average layperson could not touch anything in the temple, even a piece of furniture, because they would be consecrated and contaminated with holiness. Haran notes that this holiness cannot be removed from someone after they have come in contact, and they must then be treated the same way as all objects within the tabernacle. Finally, one potential outcome of contracting holiness is immediate death. If the death is not immediate, as seen in the many instances above, it is slow and painful.[26]

The Ark of the Covenant and the tabernacle are usually associated with the presence of Yahweh. Biblical scholars even note that they symbolize holiness and foreshadow Jesus Christ's atonement of sins. However, this seems hogwash when viewed as a concrete object, such as an electrical capacitor or advanced holographic telecommunications device.

The Ark and Egyptian Pyramids

Christopher Dunn, a career engineer and master craftsman, studied the Pyramid of Giza in depth. He analyzed the building materials and various types of stone used and theorized that the Great Pyramid was an ancient form of a power plant. His theory uses the energetic grid of the earth as a source of power for the pyramid.[27] Other theories state that water from the Nile River, located beside the pyramid complex, flowed into the lower chambers, causing a hydroelectric effect and generating power. These would both be considered old and outdated at the time of the 18th Dynasty, causing other means of power generation to be needed. The king's chamber of the Great Pyramid houses a small box, often misnamed a sarcophagus, whose dimensions match exactly those of the Ark of the Covenant. This means that the Ark could have been the power generator of the Great Pyramid, and when Moses took the Ark with him out of Egypt, he took the very thing that gave the people power.

Two Functions of the Priestly Garments

The priestly garments that Yahweh instructed Moses to create also give evidence of advanced technology and showed the real purpose of the ark. While reading through the Old Testament, it becomes alarmingly clear that essential details are provided in lengthy descriptions. Yahweh gives Moses specific instructions for the design of these garments. Much like the tabernacle and the Ark, there is more than meets the eye with the priestly garments. The garments include a breast piece, ephod, robe, coat of checker work, a turban, and a sash. Modern depictions of the garments show the Levitical priests adorned in gold, purple, blue, and fine linen as though they were a type of ancient pope to be revered. While they may have been colorful, a deeper understanding of each is necessary to understand this entire picture of advanced technology.

Exodus 28 begins the description of the garments, and Yahweh tells Moses to speak with the gifted artisans (NJKV) or the skilled workers (NIV). Biglino notes that "gifted artisans" was a mistranslation of the original Hebrew phrases. This mistranslation causes the ephod to be considered a beauty piece for the priest. The accurate translation would be more similar to the skilled workers' understanding that the ephod is created by a technician and not for beauty.[28] Madden Jones agrees with this and notes that the garments were designed for a practical function, not for beauty.[29]

The fabric is highly important to note because scholars believe each thread of blue, scarlet, purple, and linen is intertwined with gold wire on the ephod. Josephus states the importance of the order in which each of these garments were to be placed on the priest. The fine linen undergarments were followed by:

1. the linen coat of checker work made of double flax and sat close to the body, long sleeves tied fast to the arms, and reached the feet. The entire "coat" or vestment, as Josephus calls it, was snug, and no part was hanging off the priest.
2. The sash or belt was tied and taught around the waist.
3. The cap or turban was thick swathes of blue linen, doubled multiple times, and then sown. It was snug and covered the forehead.
4. The blue robe reached the priest's feet and was one piece of fabric with an opening specifically for the neck, hands, and ephod. At the bottom of the robe were bells with golden pomegranates between each bell.
5. The ephod resembled an embroidered short coat uncovered around the breast piece.[30]
    a. The ephod was also interwoven with gold and, as noted earlier, was created by technicians for a practical function. Madden Jones notes that the linen would have been wool with gold interwoven through each thread. This would have looked similar to a wire cage and could have acted as a Faraday cage known today.[31]
6. The breastplate is the last piece placed into the ephod snuggly so it does not move. The two onyx stones enclosed in fine gold wire sit on the shoulders of the ephod. Biglino recreates this ephod and notes that below the stones lay the rings of gold

> attached to the chains of twisted cord.[32] This chain then connects to the breastplate, set with twelve stones. Josephus also notes that two sets of rings were made to secure the breastplate to the ephod due to the breastplate's weight.[33]

The breastplate held twelve stones representing the twelve tribes of Israel, and each stone was named after a tribe. Two stones, Urim and Thummim, are particularly interesting because they are unknown. Madden Jones notes that Urim is translated from Hebrew *Uwriym,* meaning 'fiery' or 'lights.' The plural of *uwr* means 'fire,' 'flame,' or 'light' and comes from the root word *owr,* meaning 'to be luminous or to make luminous.' Thummim in Hebrew means 'perfection' or 'completeness.' This idea of completeness is older than that of the Hebrews and comes from the Sumerian idea of completing the connection between heaven and earth.[34]

While the Hebrew language does not provide an etymology for Urim and Thummim, the Sumerian language suggests a possible meaning. In Sumerian, *Urim* (*U'rim*) becomes *u* (height) and *rim* (reduce or shorten), and Thummim becomes *tum* (bring) and *min* (Shamash or 20).[j] Urim becomes a 'distance–shortener,' and Thummim becomes 'Shamash (or Yahweh) bringer.' Showing the connection that the ephod was used to communicate with Yahweh.[35] Josephus comments that all twelve stones stopped shining only two hundred years before he wrote *Antiquities of the Jews.*[36] The breastplate was placed around the waist by a linen girdle containing thin golden wires.

The purpose of the ephod and breastplate is shown in 1 Samuel 30:7-8 when David tells Abiathar, "Bring me the ephod." So Abiathar brought the ephod to David and David was able to speak with the Lord. In other words, David can speak with Yahweh only after he has the ephod. This means it is the central communication device between the one wearing

---

j Elohim or Anunnaki hierarchy was based on a numeric system; the number 20 was the number of Shamash. See chapter twenty.

the ephod and Yahweh. The ephod was used as a form of verbal communication after the tabernacle was no longer used.[37]

The various sightings of mysterious clouds of fire and smoke to radiation poisoning, death by poisonous gas and electrocution, and holographic communications with a craft in the vicinity of the camp all seem too fantastical when thinking about the boring old books of the Old Testament. It is precisely in these boring books that the incredibly advanced technology and extraterrestrial aspects are directly spoken about. One only needs to see past the mumbo jumbo of holiness and glory to see the truth about UFOs and holograms.

Giants, Nephilim, Watchers, or Anunnaki Hybrids? All of the Above

An Overview

"The Nephilim were on the earth in those days, and also afterward when the sons of God came to the daughters of man, and they bore children to them. These were the mighty men of old, men of renown." - Genesis 6:4 ASV

Chapter one discussed the existence of giants, chapter four discussed the sages and the Shining Ones, and chapter seven referenced the various similarities between the stories in *The Book of Enoch* and *The Lost Book of Enki*. This chapter will focus entirely on the nephilim and the existence of giants in the past. The story of the giants in the Bible begins in Genesis 6, and it is the first time that critical information is given.

"The sons of God saw that the daughters of man were attractive, and they took as their wives any they chose." – Genesis 6:2 ESV

- Sons of God – [בְּנֵי־] *B'nē* [הָאֱלֹהִים] *hā-Elōhim.*[1] This is a slight mistranslation, however.
  - Sons = [בְּנֵי־] *B'nē*
  - The = [הָ] *hā*
  - Elohim = [אֱל] *ōhim* [הִים] *El* – shining ones[a]
- The full and correct translation would be 'Sons of the Shining Ones.'

"The Nephilim were on the earth in those days, and also afterwards." The biblical author is once again, as mentioned in chapter six, giving the summarized version of history to an audience that would have been very familiar with the history of the giants. Canaanite texts note that the term "sons of god" was a regularly used term by pagans to refer to the pantheon of gods.[2]

The Sumerian story of *The Lost Book of Enki* is about the 200 Igigi who came to Earth. Their leader, Shamgaz, chose a wife from the Adapite females and started families. The story notes that they arrived on the Landing Platform in the Cedar Mountains.[3] Sitchin notes that the reference to "Mighty Men" were men who were appointed by the gods to rule in their place after the second Pyramid War.[4]

Tracing the Story from Cuneiform Text

The Sumerian version speaks of the Cedar Mountains just as the *Epic of Gilgamesh* notes the Cedar Mountains and the Landing Place.[5] Sitchin notes that the only place the Cedar Mountains could have been referring to in the Middle East would have been the Mountains of Lebanon, and the "Landing Platform" mentioned is in Baalbek.[6] This Landing Platform can be envisioned when looking at the base of the Temple of Jupiter or the Stone of the Pregnant Woman, which weighs 1,500 tons.

---

[a] Sumerian *El*: brightness or shining. Akkadian *Ilu*: the bright one. Babylonian *Ellu*: the shining one. Old Welsh *Ellyl*: shining being. Od Irish *Aillil*: shining. Anglo-Saxon *Elf*: a shining being. Old Cornish *El*: angel (O'Brien. *The Genius of the Few*, p. 27).

The Greek version of the *Book of Enoch* notes that the summit of Mount Hermon is in Lebanon.[7] O'Brien notes that Enoch's location was the summit of Mount Hermon, and Lebanon is close to the western boundary of Hermon.[8]

Marduk and the other Igigi and lower-level Anunnaki came down from heaven and started families with the Adapite females. Yahweh, like Marduk, was a lower-level Igigi, and more similarities between these two characters will become apparent in later chapters. Marduk married Sarpanit, daughter of Enmeduranki (Enoch),[9] and the Igigi lay with daughters of man and started families. This enraged Enlil, as *The Lost Book of Enki* puts it, because he saw it as abandoning their mission on Earth. Even Enki mated with an Adapite female who bore him Ziusudra.[10]

Enlil, Enki, and Ninmah were informed of the changes that were to come on earth and were falsely told by a messenger of Anu that a flood was to come and to seek refuge. Only the Anunnaki and Igigi were to be saved, and all the others were to drown. They were given a choice: return to Nibiru without their families or await the earth's changes in a spaceship. Some left, and others waited for changes in mountain tops, caves, or boats (Ziusudra). After Enlil learns of the survival of the nephilim, he is initially angry, yet the Anunnaki agrees to continue forward. They would allow the nephilim, called *lugal* (mighty-men) in cuneiform stories, to become the first rulers when kingship was handed down from the gods to man.[11]

THE WATCHERS AND GIANTS

This anger of Enlil is falsely attributed to Yahweh in the *Book of Enoch* when he learns that the angels, or Watchers, chose wives and started families. Chapter seven notes that the offspring of the Watchers are giants:

"[the women] became pregnant, and they bare great giants, whose height was three thousand ells: Who consumed all other acquisitions of men. And when men could no longer sustain them, the giants turned against them and devoured mankind." – Enoch 7:2-4.

The only time the Watchers are referenced in the Old Testament is:

"I saw in the visions of my head as I lay in bed, and behold, a watcher, a holy one, came down from heaven." – Daniel 4:13

The Hebrew word for Watchers in this verse is *Eyrim,* and they are noted as the sons of the Elohim, the inferior group, or the secondary group of the Elohim.[12] Strong's Concordance shows the word used is: (עִיר) *ir*, Aramaic root meaning 'waking' or 'wakeful.'

This description of the Watchers aligns with the description of the Igigi. They were a part of the Elohim. In *The Book of The Secrets of Enoch*, Enoch describes the Watchers to help "Methuselah and those who will come after" recognize them. He describes them as very tall men, taller than he has ever seen before, whose faces shone like the sun, eyes were like burning lamps, fire came from their lips, and their hands were whiter than snow.[13] This is the perfect description of the Shining Ones. The Watchers are even referred to as the third order of the Elohim. The first order is that of the Anunnaki, the second is the Igigi who rebelled, and the third is the Watchers whose transgressions and mess-ups could have been erased by the flood.[14]

The word "Watchers" comes from the Greek word *egrēgoroi* and can be translated as 'those who watch' or 'those who are awake.' O'Brien notes that the Watchers were chosen over the second translation because ;those who are awake' could be confused as 'those who never sleep,' and as all accounts show, the Elohim or Anunnaki are just like humans.[15] The names of the leaders of these egrēgoroi are given in Enoch 6:7: Samlazaz, Araklba, Rameel, Kokablel, Tamlel, Ramlel, Danel, Ezequiel, Baraqujal, Armaros, Batarel, Ananel, Zaqiel, Satarel, Turel, Jomjael, Sariel, and Semjaza, the leader over the 200. El runs through most of the names, showing that these, too, had the same characteristics as the Elohim, as noted by Enoch in his *The Book of the Secrets of Enoch.*

The Book of Giants gives insight into the story of the Watchers or egrēgoroi and is another bridge between the Apocrypha books and the Old Testament. It is a compilation of fragmented texts of the Hebrew story, the Ethiopic version, Greek fragments, and excerpts from Byzantine chronographer George Syncellus. It has been published in at least six or seven languages, causing confusion between the names of the giants.

This ancient text gives insight into the Watchers before they arrived on Mount Hermon. There was a war where four hundred thousand egrēgoroi were killed or imprisoned, and 200 escaped. These 200 egrēgoroi mated with the women and birthed giants.[16] Four angels descended to earth, and when the egrēgoroi discovered they were to be forcibly removed from their families, the egrēgoroi took on the shape of men. The angels separated the men from the giants, then forcibly separated the giants from their families, placing half to the east and the other half to the west. Thirty-two towns had been prepared for them. Another translation notes that 36 towns were prepared for the sons of the giants to live there because they lived a thousand years.[17]

This incredible age is seen in Og, King of Bashan, who lived 5000 years and survived the Deluge due to his giant size.[18] The story of King Og and his ultimate defeat and death is spoken of in Deuteronomy. He was the last of the Rephaim whose "bed was a bed of iron...nine cubits (12.8 feet) was its length, and four cubits (5.6 feet) its breadth" (Deut 3:11). While *The Book of Giants* is short and heavily fragmented, it gives valuable insight into the relation between the giants and nephilim or Watchers.

Origin of Nephilim

In his work, Professor Michael Heiser notes the true root of the word 'nephilim' is *yarad*, a verb relating to the act of descending. Similarly, Professor Ronald S. Hendel's work parallels this root meaning and passages in Ezekiel 32 describing the voluntary descent of heavenly warriors. It concludes that 'nephilim' incorporates both deliberate descent and involuntary falling. As noted in the previous stories, the nephilim deliberately descended upon Mount Hermon and involuntarily fell from the grace of the Elohim or Anunnaki.[19]

Josephus remarks that the descendants of Seth remained virtuous for seven generations but fell from grace when they chose perversion. They became twice as wicked and made God their enemy by keeping company with women and begetting unjust sons. These acts of Seth's lineage resembled those of the Grecians called giants. Noah attempted to help them change their ways before the flood, but his attempts fell on deaf ears.[20]

It is interesting to note that the Watchers, like men, fell because they gave knowledge to their wives and children. The *Book of Enoch* clarifies the various forms of knowledge that the Watchers gave mankind and Enoch. They were dammed because they gave man this knowledge. They were even given similar "punishments" to what men were given because of the advanced knowledge that they shared. *The Book of Giants* clearly explains this fall from grace and its reason:

"He is...with the wrath...and rebellion..., when malice and wrath arose in his camp, namely the Egrēgoroi of Heaven who in his watch-district (rebelled and) descended to the earth. They did all deeds of malice. They revealed the arts in the world and the mysteries of heaven to men. Rebellion and ruin came about the earth..." – Kephalaia, 9224-31[21]

After the flood occurred, the names of the nephilim or Watchers changed in the Old Testament, and they are referred to in four ways: Anak or Anakim, Rephaim, Emim, and Zamzummim. The Old Testament begins with stories about the giants in Numbers 13 when Moses sends a team of spies to gather intelligence on Canaan. The spies arrived in Hebron and came upon Ahiman, Sheshai, and Talmai, the descendants of Anak.[b] When they returned, they informed Moses:

"The people who dwell in the land are strong, and the cities are fortified and very large. And besides, we saw the descendants of Anak there... We are not able to go up against the people, for they are stronger than we are...So they brought to the people of Israel a bad report of the land that they had spied out, saying, 'The land, through which we had gone to spy it out, is a land that devours its inhabitants, and all the people that we saw in it are of great height. And there we saw the Nephilim, (the sons of Anak, who come from the Nephilim), and we seemed to ourselves like grasshoppers, and so we seemed to them'." – Numbers 13:28, 31-33.

SIMILARITIES FOUND AROUND THE WORLD

The idea of giants devouring men is not new, as it draws upon the stories of the Navajo and Choctaw tribes and the red-haired white cannibal giants. The Iranian and Middle Persians translate giant as 'monster,' 'gigantic,' or 'monstrous.' This correlates with Josephus and the fact that the giants did whatever they wanted because of their confidence in their size.

While the Israelites wander through the wilderness, they encounter multiple groups of giants. The writer of Deuteronomy notes that Moab was once the place where:

"The Emim formerly lived there, a people great and many, as tall as the Anakim. Like the Anakim, they are also counted as Rephaim, but the Moabites call them Emim." – Deuteronomy 2:10-11

The writer also claims that in the land of Moab:

b Numbers 13:22

. . .

"It is also counted as the land of Rephaim, Rephaim formerly lived there- but the Ammonites call them Zamzummim- a people great and many, and tall as the Anakim, but the Lord destroyed them before the Ammonites." – Deuteronomy 2:20-21

The Anakim or Anak are also referenced in *The Lost Book of Enki* and Enki precisely notes the definition of Anak as "Anunnaki-made,"[22] even if the spelling had not already made it abundantly clear when looking at the various stories side by side. After the Emim and Zamzummim were destroyed, the Anakim were displaced. The writer of Joshua writes:

"And Joshua came at that time and cut off the Anakim from the hill country, from Hebron, from Debir, from Anad, and from all the hill country of Judah, and all the hill country of Israel. Joshua devoted them to the destruction of their cities. There was none of the Anakim left in the land of the people of Israel. Only in Gaza, Gath, and Ashdod did some remain." – Joshua 11:21-22

This would be it if there were ever a story of intentional stamping out of a group of people or giants. It is as if Joshua made it his mission to destroy the giants from the region that Yahweh promised his people.

The three leaders of the Anakim (Sheshai, Ahiman, and Talmai) were driven out by Caleb in Hebron,[c] but it was Judah who ultimately defeated them.[d] 1 Samuel 17 notes that the remaining giants were in Philistia and their "champion Goliath of Gath, whose height was six cubits and a span,"[e] or almost nine feet tall.[23] 2 Samuel 21 notes the war between David and the Philistines. Further confirmation that the giants were still in existence is the descriptions and names of the Philistines:

---

c Joshua 14:9
d Judges 1:8-10
e 1 Samuel 17:4

"Ishbi-benob, one of the descendants of the giants, whose spear weighed three hundred shekels of bronze (about 3,300 grams),"[f] and "Saph, one of the descendants of the giants, Goliath the Gittite."[g] There was another war in Gath and a "man of great stature, who had six fingers on each hand, and six toes on each foot, twenty-four in number, and he also was descended from the giants."[h] The validity of the David and Goliath story, along with David's true identity, will be covered in the next chapter.

The fossil records discussed in chapter two now have evidence of their existence in ancient history. While there may be muddied waters and information lost in translation over the millennia, the information examined in this chapter clarifies who the Anunnaki, Igigi, Elohim, nephilim, and giants were. The Anunnaki/Elohim were the leading group and the Igigi were the second group of Anunnaki/Elohim that did the work of the gods and rebelled, then came down and mated with the Adapite females. This mating birthed the third group of Anunnaki, nephilim, giants, or egrēgoroi. The giants were monstrous beings who ate the Adapite males and females and were either wiped out by the flood or intentionally stamped out by Yahweh, according to the stories of the Old Testament.

Lastly, it would be advantageous to remember the connection between the giants or nephilim and their true origins. *Brown Driver Briggs Hebrew and English Lexicon* translates *Anak*[i] as "long-necked (tall) men, early giant people about Hebron and Philistia" and translates *nephilim* as 'giants.' The etymology of *Nephila*[j] (singular) is of Aramaic origin and translates to 'Orion.' Biglino notes that the term *nephilim* is the intersection of Greek, Aramaic, and Hebrew mythology.[24] Could this be the cultural remembrance of the origins of the gods with the stories of the Pleiades, Sirius, and Orion?

---

f 2 Samuel 21:16

g 2 Samuel 21:18-19

h 2 Samuel 21:20, 1 Chronicles 20:6

i Strong's Concordance No. 6061.

j Strong's Concordance No. 5303.

KING DAVID

In chapter eleven, it was shown that Hebrew emerged sometime within the 10th century BCE with origins from the Phoenician script, so the books of the Old Testament could have only been written after this time. Osman points out that the Biblical writers wrote historically accurate accounts of famous people but changed their geography and timelines.[1] Chapters nine and ten showed the history of the 18th Egyptian Dynasty and the drastic changes from the new monotheistic religion following Aten/Amen Ra. Speaking about the Amarna kings after the beginning of the 19th Dynasty was forbidden. Therefore, names needed to be changed and timelines altered to tell stories about the new god of Israel and His chosen people.

This idea of intermingling events, names, and dates causes some retroactive understanding of why Osman consistently claims that both *The Secret Origin of the Bible's Royal Bloodline* and *The Egyptian Origins of King David and the Temple of Solomon* show there is no archaeological evidence for the existence of King David or King Solomon's vast empire. Osman argues that King David is Pharaoh Thutmose III and King Solomon is Pharaoh Amenhotep III, father of Akhenaten. Examining each story's complexities can reveal a piece of truth and understanding.

Biblical Story from Moses to David

After the death of Moses, the Israelites wandered through the wilderness and finally settled in the Promised Land. Osman calls the group of judges who ruled over the Israelites during this time a loose confederation. The last judge within this loose confederation is Samuel. During his lifetime, the Israelites came under attack from the Philistines. The fortified cities of Ashdod, Gaza, Gath, Ashkelon, and Ekron lied on the southeast shore of the Mediterranean Sea between Tel Aviv and the Gaza Strip. After the Philistines defeated the Israelites, they brought the Ark into their camp to help them in battle. This attempt at victory was an immediate blow as the Philistines attacked them again and then took the Ark with them.

To face the constant threat of the Philistines, the Israelites demanded that they become a nation with a king. Samuel anointed Saul, the son of Kish the Benjamite, to be the first king of Israel. Saul loved music and needed a personal harp player. David was a 15 year old boy harpist who looked after his father's sheep. Saul learned about David's love of playing the harp and took him in as his harpist (1 Samuel 16:14-23).

An interesting side note exists before this story when Yahweh speaks to the prophet Samuel:

"I regret that I have made Saul king, for he has turned back from following me and has not performed my commandments." – 1 Samuel 15:11 ESV

Samuel then tells Saul that Yahweh is not happy with him as king because Saul listens to the people, not Yahweh. Saul attempts to rectify this situation, but it is too late and he is smote. Samuel grieves over Saul until his death in 1 Samuel 15:35. Yahweh speaks to Samuel and tells him that he has found a new king. Jesse brings six of his seven sons to meet with Samuel while David is still with the sheep. Samuel sends for David, and upon his arrival, Samuel notes:

. . .

"He was ruddy, had beautiful eyes, and was handsome. And the Lord said, 'Arise, anoint him, for this is he.' Then Samuel took the horn of oil and anointed him among his brothers. And the Spirit of the Lord rushed upon David from that day forward." – 1 Samuel 16:12-13 ESV

When it came time for Saul to lead the Israelites into battle against the Philistines, Goliath met them along the western edge of the Judah hills. He challenged the Israelites to bring forth their champion and fight him. The winner of this single fight would decide the outcome of the battle between the Philistines and the Israelites. Chance would have it that David came to bring food to his elder brothers on the battlefield. He heard that no one would answer Goliath's challenge and decided, with no prior training or military understanding, that he would fight against the giant covered in bronze armor.

Armed with a staff, sling, five smooth stones, and childlike unquestionable faith in Yahweh, David placed a stone in his sling, aimed, and hit the giant in the center of the forehead. Goliath fell to the ground, and in an act that could only be described as brutal, the music-loving harpist cut off the head of the giant with a sword that came out of nowhere. This act from a teenage boy struck fear into the hearts of the Philistines, and they fled the battlefield. After the defeat of Goliath, David was praised and made commander of Saul's armies. David became the people's hero, and Saul became jealous and placed David in situations where David may die. He offered his daughter, Michal, but under the condition that David bring him foreskins of 100 Philistines. David returned with the foreskins of 200 Philistines.

David then learned of Saul's plan to kill him and fled with 600 supporters to join Achish, King of Gath. Sixteen months later, fighting began again in Gath between the Philistines and Israelites, which resulted in the death of Saul and three of his sons. David then went to Judah with his 600 followers and was anointed King of Judah at the age of 30. Following Saul's death, his surviving son Ish-bosheth became

King of Israel. War then broke out between Israel and Judah and ended after Ish-bosheth was killed by two of his captains. David united the two separate kingdoms and conquered Jerusalem's fortress.

David built his house in Jerusalem with the help of Hiram, King of Tyre in Phoenicia, and called it the City of David. He went with 30,000 of his men to Gibeah to bring the Ark back to Jerusalem and kept it north of the city in a tabernacle on Mount Moriah. David finally defeated the Philistines and quickly grew his empire by conquering surrounding territories. After David established an empire between the Euphrates River in northern Syria and the Nile River in Egypt, he divided the empire into 12 distinct districts. Each district had military, civil, and religious organizations. David had but one great weakness: his lustful passion for Bathsheba.

He was on his balcony one night when he discovered Bathsheba bathing. He learned that she was the wife of Uriah, one of David's men fighting at Rabbah. David slept with Bathsheba and she became pregnant. Upon learning of this, David attempted to cover up his affair and have Uriah return home and sleep with his wife. When this plan failed, David ordered Uriah to be placed at the front of the battlefield and be killed. David then married Bathsheba, and the child born out of sin died. She later conceived Solomon, who became the most important of David's sons and followed him in the line of succession.

A second variation in David's character occurs with the confrontation with Absalom. Absalom attempted to overthrow his father and rule the kingdom after his sister Tamar was raped by his half-brother Ammon. Absalom was upset that David did nothing and left Jerusalem for Hebron. He declared himself king. David learned that Absalom was coming to take power over Jerusalem with the support of Israel and Judah so fled east of Jordan with 600 men. Absalom took over the City of David without challenge and, with his army, pursued David. The two armies met, and Absalom, along with his army, was defeated and he was killed.

David returned to Jerusalem and ruled his empire until he was 70 years old. His son Adonijah attempted to conspire against David and declare himself king. Bathsheba stepped in and persuaded David to name Solomon king while he was still alive. Solomon ruled the empire with an army of 2000 horses and men and 1,400 chariots. After Solomon's death, this once great empire completely vanished off the face of the earth.

Archaeological Evidence of David's Empire

Archaeological evidence shows that the Israelites were a group of individual tribes involved in various conflicts over a long period. This directly contradicts the story found in the Old Testament that Joshua conquered Canaan towards the end of the 13th century.

Returning to Egyptian history, the region of Gezer, known as Canaan, was controlled until the end of Rameses III's rule and the subsequent succession of Rameses IV.[2] Archaeological evidence shows that the Israelites started establishing new settlements in Dan only after Egypt lost control over Canaan (known as Habiru or Hyksos from chapter eight).[3] The Philistines invaded this region and eventually took over Canaan, naming their land Palestine. After the Philistines established Palestine in the second half of the 12th century BCE, they began expanding their empire into the same land Saul and David attempted to take over.[4] This is where the beginning of the story of David begins—or is it?

This summary of David's life and rise to power shows variations in his character. How is it that a shepherd who loves music and plays the harp is a military genius who rises to the commander level after one victory without prior training? Why would a man who defeated the giants that lived in Philistia flee because King Saul had created a plan to kill him? While a case can be made for David and his fallible humanness, another more concrete case lurks under the story: David is a combination of two people.

Israeli biblical scholar Moshe Garsiel studied the *Book of Samuel* to find out why there are so many discrepancies within the story. The Book of Samuel is a continuation of Genesis to Judges. Multiple sources of oral transmission of stories cause the historical accuracy of events to become intermingled with other parallel narratives that contain different theological and social conceptions. This causes the historical accuracy of the stories to be lost in time, and the theological significance of the story becomes more important than the historical evidence based on archaeological discoveries.[5]

The major problem with this entire view of scripture is that the Bible is supposed to be the infallible word of God. Once again, the "divine word of god" must be set aside to uncover the truth behind the stories and their discrepancies. Osman does this and shows how Moshe Garsiel's explanation of 1 Samuel blends fragmented stories, local hero sagas, and archived historical documents.[6] He recounts Julius Wellhausen's work, which suggests two main sources are woven together. The first is the accurate account, and the second contains theological meanings.[7]

Osman points to this problem and concludes that the authors and editors of 1 and 2 Samuel were more interested in representing David as a religious figure than in portraying the historical data and accuracy of his life.[8] Philip R. Davies brings this sentiment home when he comments about the lack of archaeological evidence against Biblical accounts that causes the historical figure of King David to equal that of King Arthur.[9] It would be a bold claim if the possibility of truth were absent. Israel Finkelstein echoes this sentiment and refers to the Biblical account of David and Solomon as a makeshift collection of folklore.[10]

The archeological evidence shows that King David's empire covered the land from the Nile to the Euphrates. While the land mass is vast, the Bible says he only had an army of 600 men which is an incongruency. The Bible also says King David lived during the 10th century BCE. Yet, the only archaeological evidence of empires in this region is that of Thutmose III in the 15th century BCE and Cyrus the Great of Persia after he conquered Egypt in the 6th century BCE.

Osman claims that three different stories of David have been combined to hide the actual historical character. The first story is that of David, a tribal chief with a small army of only 600 men. To create the jump from the first to the second story, the story of David and Goliath was created to help understand why he was exalted from the small tribal chief to the second story of David, King David, who rules over an empire and was chosen by Yahweh. The third story is created to explain Solomon as the successor of David and Bathsheba.[11]

Israel Finkelstein explains the capture of Jerusalem and the City of David by noting that archaeologists, in their search for evidence, identified the region now known as Ophel along the steep ridge outside the walls of the Ottoman city. He explained that while traces of pottery and fragments have been found in this city, using the biblical narrative as a reliable source for excavation is a type of circular reasoning.[12]

This circular reasoning is common when attempting to find archaeological evidence of King David's and Solomon's Empire. Finkelstein closes the case on the idea of the existence of this great empire, noting that the archaeological evidence found within Jerusalem shows that during the life of David and Solomon, the city was nothing more than a small village about three to four acres in size.[13] The question then becomes, who is the king who ruled over a vast empire?

KING DAVID AS THUTMOSE III

Looking back, the evidence of the grand empire attributed to King David is found during the time of Pharaoh Thutmose III. This creates the first part of Wellhausen's argument of two parallel sources being woven together to create one cohesive story. The second part is of the tribal chief who lived five centuries later. The biblical scribes did this to conceal the relationship between the pharaohs of Egypt and the tribes of Israel. Oddly, however, Egyptian sources such as the Amarna letters openly described this relationship. When scholars looked through the Amarna letters, the Hebrew people of the Old Testament were quickly identified as Hapiru, a term denoting the outcasts within their society.[14]

Before fully splitting these two stories to find the truth, one source shows the same title for two different kings. David is known as the warrior king. He rose from shepherd to commander within Saul's army and eventually became king over all of Israel. He became this warrior king by honoring Yahweh in all his military battles. The buildings erected at Karnak by Thutmose III give the impression that this pharaoh was a warrior king whose victories concurrently honored the king and the state god Amun.[15] The stories told within the temple walls of Karnak were copied from a scribe at Thutmose III's side during the events.[16]

The temple walls and columns show how Thutmose III's real story became adapted to fit the Hebrew narrative. It details Thutmose III's military campaign against Megiddo, including the seven-month siege, the escape of the king, leaving the battlefield for a brief time, and his ultimate victory. This gives historical evidence for David's battle against Rabbah, detailed in 2 Samuel 10-11. The temple shows Egypt taking control over Qadesh and over a thousand names of Canaanite cities during the time of Thutmose III. This link between Thutmose III and King David is found in Nehemiah 11:1 when Jerusalem is called "the holy city." The Hebrew translation of the holy city is *Yerushalayim 'ir ha-qodesh*[a]. The Hebrew '*ha-Qudesh*' is translated into the Arabic *al-Quds,* meaning 'Jerusalem.'[17] This is the origin of Zion and the City of David found at the Karnak Temple.

DAVID AND GOLIATH

The extraordinary story of David defeating Goliath the giant varies greatly depending on the version of the Bible read. In the Septuagint, Masoretic text, and modern-day translations of 1 Samuel 17, there are so many different forms of one story that the truth comes into question. The David and Goliath story is again a Hebrew retelling of an Egyptian story from the 12th Dynasty of a courtier named Sinuhe. Osman notes that *The Autobiography of Sinuhe* was added to David's story to give a young harpist shepherd the bravery and courage needed to become a great king one day.

Sinuhe was the courtier of Neferu, daughter of the founder of the 12th Dynasty, Amenemhat I. This story takes place around 1960 BCE, during the last year of Amenemhat's 30 year reign. Sinuhe was with the

---

[a] Strong's Concordance notes that it is translated '*birusalim 'ir haq qōdeš*' with *š* becoming *sh*.

Egyptian army, led by Sesostris, eldest son of the pharaoh, when they returned from Libyan campaigns. Messages were sent to all the pharaoh's sons about an attempt on the pharaoh's life. Sinuhe fled the encampment out of fear and eventually traveled to Sinai. The chief of a Bedouin tribe gave him food and drink and helped him reach southern Canaan. Sinuhe traveled along the Way of the Sea[b] and befriended Prince Nenshi.

Nenshi saw Sinuhe as an ally because Sinuhe spoke highly of the pharaoh. Nenshi "placed [Sinuhe] in front of his children and married [him] to [Nenshi's] eldest daughter...[gave him] plentiful land... and [made him] the ruler of a tribe. [Sinuhe's] children grew into mighty men, each in control of his tribe. [Sinuhe] gave water to thirsty men... [and] rescued him who had been robbed."[18] Sinuhe also opposed the movements of the Asiatics when they "started to rebel against the rulers of the deserts."[19] Sinuhe spent many years as the commander of Nenshi's army. He was successful and found favor in the heart of Nenshi, even to the point of being placed above his children.[20]

Osman retells this story in *The Egyptian Origins of King David and the Temple of Solomon* and gives the history behind the geography during the 12th century BCE. This is because the land that Sinuhe refers to in his story can only be that of Canaan. Within Canaan, there were fortified city-states surrounded by strong walls. Sinuhe speaks of people who lived outside the walls and were seminomadic hunters and shepherds.[21] Chapter eight showed that this city was known as Avaris.

William Kelly Simpson's work references in the footnotes that this story was used as a reference for the David and Goliath story. In his notes, he references the earlier work by G. Lanczkowski, who remarks that the David and Goliath story is nothing more than a literary prototype.[22] This fits in with the idea that the entire story has been made up to bring David from a young shepherd and to create the extraordinary character of David.

---

b Around 2200 – 1550 BCE, the Way of the Sea crossed with the northern portion of the Dead Sea—this connected part of Jerusalem near the Jordan River with the Mediterranean.

King David and Thutmose III: Chosen of Yahweh or Ra

Another similarity between David and Thutmose III is their relation to Yahweh and Ra. In 1 Samuel 16:1-13, the authors tell the story of when Yahweh chose David to rule as king. Yahweh told Samuel to go to Jesse the Bethlehemite and offer a sacrifice. After the sacrifice, Yahweh would show Samuel who he has chosen as the next king. Samuel does as he is instructed. David is not present with the other sons of Jesse and is called upon by Samuel. After he arrives from tending to the sheep, Samuel anoints David, then:

"The Spirit (*ruach*) *of* the Lord rushed upon David from that day forward."

– 1 Samuel 15:13 ESV

Could this possibly be another attempt to weave together two separate stories to create the heroic nature of King David?

Chapter nine noted that Thutmose III was the son of a concubine named Isis and not the legitimate heir to the Egyptian throne due to strict Pharaonic succession rules. This meant that in the eyes of Egyptian custom, he had no legal right to the throne. Queen Hatshepsut understood this and objected to Thutmose III becoming the next king. To ensure his son would rule, Thutmose II took his son to the Temple of Amun to have his son adopted by Amun-Ra. The inscriptions at Karnak tell the story of Thutmose III becoming the son of Ra.

The story begins with a sacrifice, just like David becoming anointed by Samuel. For the sake of clarity, the information presented in Chapter nine is repeated here, as it will be necessary for the comprehension of this story:

139. "His majesty [Thutmose III] placed for him incense upon the fire,

and offered him a great oblation consisting of oxen, calves, mountain goats.

140. On recognizing me, lo, [Amun-Ra] halted ---- [I threw myself on] the pavement, I prostrated myself in his presence. He set me before his majesty...

141. [He opened for] me the doors of heaven; he opened the portals of the horizon of Rd. I flew to heaven as a divine hawk, beholding his form in heaven; I adorned his majesty... feast. I saw the glorious forms of the Horizon-God upon his mysterious ways in heaven.

142. Ra himself established me. I was dignified with the diadems which [we]re upon his head, his serpent-diadem, rested upon [my forehead] ... [he satisfied] me with all his glories; I was sated with the counsels of the gods, like Horus, when he counted his body at the house of my father, Amon-Ra. I was [present]ed with the dignities of a god, with ... my diadems.

143. His own titulary was affixed for me.

146. ... [in this my name]. King of Upper and Lower Egypt, Lord of the Two Lands: "Menkheperre" (the being of Ra abides)."[23]

The inscriptions at Karnak were written during Thutmose III's life, not centuries later. Osman follows the trail of how Thutmose III's name was changed due to the changing of languages throughout the centuries. The hieroglyphic writing of Thutmose's name was broken into two parts: *tut* or *tw* and *mos*. *Tut* is transliterated to *Dwd* (beloved) in Hebrew, which translates to *David*, and *mos* means 'child' or 'son.'[24]

The *Encyclopaedia Judaica* notes that David was not even the character's original name; it was Elhanan, which was changed once he became king.[25] This is mentioned in 2 Samuel 21:19: "Elhanan slew the brother of Goliath the Gittite" and causes scholars to backpedal and make sense of the two names.[26] The name change from Elhanan to David is found in the Targum, the early Hebrew Bible translated to Aramaic.[27] This second puzzle piece shows that the character of King David is nothing but fiction.

David, Bathsheba, and Solomon

Osman completes his argument for David as Thutmose III with the story of David and Bathsheba. As already noted in chapter nine, Isaac was not the son of Abraham but was the son of a Pharaoh, Thutmose III.

Yes, the mental gymnastics of the original biblical narrators must have been incredible.

Rabbi Moshe Garsiel has already remarked that the cyclical nature of the books of Samuel is similar to that of the previous books due to the parallel source narratives. Or could this have all been done for a purpose?

While 2 Samuel 11 tells the story of David and Bathsheba, Osman notes that this passage with 2 Samuel 12:1-25 is an insertion into the previous chapter that speaks about the battle with the Ammonites after the original story was written. How does David, a warrior king, stay home while the rest of his army is in the middle of a battle? Edwin Good discovered the ironic quality of this story considering the custom of kings fighting alongside their armies.[28]

Osman compares Abraham to Uriah and notes that they are both foreigners whose wives married a pharaoh (king) and became pregnant. The wives gave birth to sons, and both sons must die because they are born out of sin. Isaac's life is spared, but the child from David and Bathsheba must die. The Genesis story removes the pharaoh when he learns that Sarah is married and sends them away. In the David story, David gets rid of Uriah and is sent to his death. This is two different points of view for one story.

In both stories, the ruler is threatened due to his actions, and only the woman escapes the punishment. Osman breaks down the name of

Uriah into two parts, *Ur* and *yah*. *Ur* means *city* or *light* and relates to a location in northern Mesopotamia (remembering also that Yahweh called Abram to travel from Ur to the land of Canaan), and *yah* is the shortening of Yahweh. The exact name breakdown can be seen in the name of Bathsheba. He breaks Bathsheba into two parts *Bath* and *sheba*. *Bath* means *girl* or *daughter*, and *Sheba* refers to the land south of Canaan named Beersheba. Therefore, *Uriah* means 'light of Yahweh,' who came from Ur, the land of the Chaldeans, and Bathsheba means daughter of Beersheba, where Sarah and Abraham settled after they returned from Egypt.[29]

This is the third piece of the puzzle, ultimately creating a heroic man whom Yahweh chose to establish Israel as a nation. This vast change helped hide the truth behind the Amarna kings and their transformation of Egypt and elevate the Hebrew story.

The Real Solomon?

Like everything else in this book so far, the story of Solomon is much more than meets the eye. The truth is hard to decipher from a made-up empire to twisted stories, deceit, intentional reframing, symbolism, and changing a narrative. The foundation of a lie always holds a speck of truth. Sifting and sorting information from every angle will unveil the truth about the monumental character known as Solomon. Until the big secret keepers reveal the truth, it is up to everyone else to find it, even when it becomes uncomfortable.

Solomon and Amenhotep III

Osman greatly served the world when he published multiple books about Egyptian pharaohs and biblical patriarchs. Whether or not he tells the actual story is beside the point. He gives a new perspective on old characters and questions the validity of the Old Testament and its patriarchs. Many similarities coincidently align between Amenhotep III and Solomon, including the type of reigns they held, marriages, titles, geographical locations, building projects, and connection to their gods. Could these all be coincidences or is something else going on?

. . .

Reigns of Peace

Amenhotep III's reign was peaceful and prosperous. It was the pinnacle of the 18th Dynasty and was obtained without significant military advancement, as in previous reigns. It is well documented in Egyptian hieroglyphics and archeological artifacts. Solomon's reign was also said to be peaceful and prosperous. Yet, there is minimal archaeological evidence to prove his empire. The story of his birth notes:

"Then David comforted his wife, Bathsheba, and went over to her and lay with her, and she bore a son, and he called his name Solomon. And the Lord loved him and sent a message to Nathan, the prophet. So, he called his name Jedidiah because of the Lord." – 2 Samuel 12:24-25 ESV

Wait, Solomon has two names. Why?

Jedidiah means 'beloved of the Lord.' Solomon or *Shelomoh* is the Hebrew spelling derived from *shalom,* which means 'peace.'[a] Instead of Jedidiah, the biblical editors used Solomon as a title when referring to this biblical character.[1] Even with the name change, research still yields results that show Solomon's kingdom, without archaeological evidence of this kingdom. Scholars state that he ruled during the 10th century BCE, and the Bible notes that his kingdom spanned from the "Euphrates River to the land of the Philistines and down to Egypt" (1 Kings 4:20). Archaeological evidence shows that Amenhotep III ruled over this region during the 14th century BCE.

Solomon's empire was named Jerusalem. As noted in the previous chapter, it was known as Qadesh, the holy city, during the reign of Thutmose III. The Amarna letters, however, show the change in name from Qadesh to Jerusalem during the reign of Amenhotep III. Jerusalem is, of course, the Arabic version of the Akkadian name *Urusalim*, meaning

---

[a] https://www.abarim-publications.com/Meaning/Solomon.html

'the foundation of peace' or 'to establish peace.'[b] In *Selection from the Tell El-Amarna Letters*, the mayor of Urusalim, Abi-Heba, sent six letters to Amenhotep III telling him that the Habiru (Hebrews)[c] were becoming a problem in Urusalim. As governor of Urusalim under the rule of Egypt, he asked Amenhotep III for assistance from Egyptian forces. He also noted that if he did not receive help, the land of Urusalim would be destroyed. [2]

The Pharaoh's Daughter

1 Kings references Solomon marrying the pharaoh's daughter five times.[d] 1 Kings 9:16 states that the pharaoh captured and destroyed Gezer with fire and killed the Canaanites. He then gave Gezer to Solomon as a dowry for his daughter.

"Solomon made a marriage alliance with Pharaoh, king of Egypt. He took Pharaoh's daughter and brought her into the city of David." – 1 Kings 3:1 ESV

The Amarna letter notes that the region of Gezer was under the control of Egypt until it was overtaken by the Apiru (Habiru) after the mayor Milkilu was threatened with his death and the kidnapping of his entire family. Within a hundred years, the land of Gezer was recaptured by Pharaoh Merneptah (the fourth king of the 19th Dynasty). Israel Stela notes that Canaan was taken captive, made nonexistent, Gezer was conquered, and Israel destroyed.[3]

While debate can be made on whether Solomon married a pharaoh's

---

[b] *Uru* means to found or establish. *Salim* is a variation of the Hebrew *shalom* and the Arabic *salam* meaning peace (Osman. *The Egyptian Origins of King David and the Temple of Solomon*, p. 103).

[c] Handcock notes the similarities between the stories of the Habiru and Hebrews in various texts and traditions (Handcock. *Selections from the Tell-El Amarna Letters*, p. 5).

[d] 1 Kings 3:1, 7:8, 9:16, 9:24, 11:1

daughter, Egyptian history tells a different story. Egyptian pharaohs were known for keeping their family relationships within the Egyptian empire. Restrictions were placed on the marriages of Egyptian princesses.[4] Amarna Letter Four gives insight into these restrictions: Kadašman-Enlil, the Babylonian king, asks why Amenhotep III stated in his previous letters that Egyptian pharaohs do not marry their daughters to anyone outside the Egyptian empire.[5] There could be an argument about a possible Egyptian princess married to Niqmaddu II (the second king of Ugarit from 1350-1313 BCE) from a relief but there is no inscription on the relief and no definitive proof.[6]

Solomon also had wives from the Moabites, Ammonites, Edomites, Sidonians, and Hittites, and had:

"...seven hundred wives of royal birth (princesses), and three hundred concubines. And his wives turned away his heart." – 1 Kings 11:3 ESV

Amenhotep III may not have had 700 wives; he married two princesses from Syria, two from Babylonia, one from Arzawa, and two from Mitanni, who arrived with 317 chambermaids.[7]

It is uncertain whether or not Solomon married the pharaoh's daughter. It is a historical fact that Amenhotep III married the daughter of Thutmose IV, his sister Sitamun, to inherit the throne and establish his rule through Egyptian succession rules. The biblical editors could have noted that Solomon married an Egyptian princess to exalt his authority over all other regional kings. [8]

Title: Son of God

The inscriptions at Karnak and Luxor make it easy to see that Amenhotep III consistently refers to himself as the Son of Re and associates himself with Aten as his son. Solomon also receives this title. 2 Samuel 7:12-14 recalls the covenant between David and Yahweh. Yahweh tells David that he will raise his son and build his temple, and says:

. . .

"I shall be to him a father, and he shall be to me a son." – 1 Kings 7:14 ESV

Anointment as King

Chapter fourteen recalls that the Israelites were nomadic people before Saul's arrival. After the reigns of Saul, David, and Solomon, Israel returned to a small group living in the Highlands of Palestine. The previous texts in the Old Testament show the Israelite beliefs and the covenant between Yahweh and the Israelites. This covenant held that the Israelites followed Yahweh with the promise that he would give them victory. However, the Israelites departed from their covenant only during the reigns of Saul, David, and Solomon. They were kings, and kings in ancient times were regarded as descendants of the gods and had a divine right to rule.

Anointing a king at the time of their coronation was an Egyptian custom. The king was anointed with oils and perfumes on a throne of gold and ivory atop six steps.[9] The Bible applies the same Egyptian customs to the anointing of Saul (1 Samuel 9:16), David (1 Samuel 16:13), and Solomon (1 Kings 1:39). The Hebrew word for anoint is *mesheh* and comes from the Egyptian *meseh,* in hieroglyphics is seen as two crocodiles because the fat from two crocodiles is used in anointing oil.[10]

Geographical Similarities and Military Structures

Amenhotep III and Solomon introduced the chariots as separate military entities during their reigns. 1 Kings 4:26 gives Solomon's account, and Egyptian history shows the introduction of the chariot branch in one of Yuya's titles, "Deputy of His Majesty in the Chariotry." Amenhotep III split Egypt into two distinct taxation and administration districts. Solomon united twelve separate taxation districts into one. [11]

As noted previously, Solomon ruled over the land from the Euphrates to the Philistines, the border of Egypt, and Tiphsah to Gaza. The biblical editors said he ruled over Hazor, Megiddo, and Gezer.

"And this is the account of the forced labor which King Solomon levied to build the house of the Lord and his own house and the Millo and the wall of Jerusalem and Hazor and Megiddo and Gezer." – 1 Kings 9:15 ESV

Remembering from the previous chapter that Thutmose III and David expanded their empire into this region, Solomon and Amenhotep III inherited this land from their predecessors. According to factual historical evidence, the Israelites inhabited the area during the reign of Jeroboam II.[12] Finkelstein, Osman, and many more biblical scholars remind everyone that using biblical narratives to search for archaeological evidence leads to a dead end.

The Temple, Temple Complex, and Kingdom of Tyre

The massive building projects included a king's palace, a house for the pharaoh's daughter, a throne room, a hall of pillars, and the House of Forest of Lebanon. 1 Kings 6-7 notes the various projects that Solomon took on, and it took him 13 years to complete them. Osman remarks that Amenhotep III built each of these projects during his reign, which took him 18 years to complete.[13]

The Solomon story notes that Hiram Ih, the King of Tyre, assisted Solomon in building the temple complex by sending Hiram Abiff. Tyre had close relations with Egypt during the 18th Dynasty, and in The Amarna Letters EA 146-155, Abimilki, governor of Tyre, sends correspondence to the Egyptian pharaoh. Wallace Flemming notes that the Kingdom of Tyre was considered invincible around 1400 BCE and had close ties to Egypt during the reign of Amenhotep IV (Akhenaten).[14] Solomon corresponded with Hiram, King of Tyre, and Amenhotep III corresponded with Abimilki, King of Tyre (named King of Tyre by Akhenaten after the death of Amenhotep III). Each ruler had a relationship with the kingdom of Tyre. Solomon's temple complex was built in Jerusalem, and Amenhotep III's was built in Thebes, also known as Qadesh, and as seen earlier in this chapter, later named Urusalim.

The Historical

To a large extent, academia questions Solomon's story's legitimacy and historical accuracy. During the reign of Solomon, Israel was a small hill country kingdom with no massive construction works, no signs of commercial prosperity, and no literacy.[15] Lemche gives insight into the realities of Israel during the 10th century BCE and remarks that life must have been difficult in the central Highlands of Palestine. The kind of surplus needed to produce a scribal class, along with the leisure time required to write historical narratives and current events, was absent during the 11th and 10th centuries BCE.[16] This then begs the question of the historical accuracy of biblical accounts and why Solomon is so highly revered.

If the biblical narrative consists of details regarding trade transactions, the unification of twelve taxation districts into one administration, and the monetary knowledge of gold and items for the temple, then the biblical editors would have first-hand knowledge of these things. In contrast, Finkelstein points to the reign of King Jehoash in the 9th century BCE. He argues that, with creative writing, the biblical editors could construct a story that brought together the traditions of both Judah and Israel.[17]

After the first ten chapters of Genesis, the stories of the great patriarchs begin to go awry. When these stories are placed next to other ancient literary texts and archaeological data, holes start to appear. Albert Pike remarks in *Morals and Dogma* that Moses taught a religion incorporating the greatest teachings the ancient world had to offer, along with the worst. This religion intentionally promoted the principle of excluding all other gods and propagating one god.[18] The priests continued this tradition, and this specific creative writing of historically inaccurate accounts helped to reform the cult of Josiah and promote Deuteronomic ideology.[19]

Furthermore, suppose the account in Chronicles is to be portrayed as a historically accurate rendition of the 10th century BCE. Why does 1 Chronicles 29:7 state, "They gave for the service of the house of God 5,000 talents and 10,000 *darics*"? Why does 2 Chronicles 36:22-23 mention Cyrus, the King of Persia, and his edict to rebuild the temple? The daric was a Persian coin first minted in 515 BCE, and Cyrus' edict took place in 539 BCE. The only logical answer is that the story of David and Solomon in Chronicles was written after 515 BCE, five centuries after the life of David.[20]

THE MYSTICAL AND FANTASTICAL

Suppose the real story of King Solomon and the Israelite Empire of Jerusalem is about a local tribe chieftain named Jedidiah who ruled over a small nomadic group in the Highlands of Palestine. Is religious manipulation the only reason for keeping this story a secret?

Multiple aspects of Solomon's story interlink with Hiram Abiff. 1 Kings and 2 Chronicles only touch upon this. The interlink between these two characters dives into the mysterious teachings of Freemasonry and Mystery teachings that chapter eighteen will dive much deeper into. Using the verse from 2 Samuel 12:24-25, the actual name of the king, Jedidiah, and the title of Solomon gives a clearer picture. Here's why: Manly P. Hall notes in Freemasonry tradition that the name Solomon is the symbolic expression of solar energy and is shown as SOL-OM-ON. The unification of the three syllables (SOL-OM-ON) brings together the teachings and mysteries of the three lights, or cults[e]—stellar, lunar, and solar.[21] Therefore, in the masonry tradition, Solomon as a title symbolizes unification and the name for the unnamable light in three languages.[22] The story then becomes a type of symbolism rather than historical fact. Just as the parables in the New Testament are meant as an

---

[e] "Cults" refers to the various formations of religion in the ancient world, as noted by Maxwell in *That Old-Time Religion*, p. 4.

allegory with hidden meaning, the figure of King Solomon is a type of Old Testament parable that could have been written to unite Judah and Israel[23] or hide ancient mystery teachings of universal enlightenment within one grand character's life.[24]

In traditions other than Christianity, Solomon was seen as a magician. That's right, magic—not the magical illusions seen today but the magic of understanding the nature of this reality. The *Nag Hamadi* book *Testament of Truth* reveals this magic: "Solomon... is one who built Jerusalem by means of demons because he received [power]."[25] Josephus also spoke about Solomon, whose wisdom was so great that it equaled that of the Egyptians[26] (whose knowledge came from Hermes [Thoth] and was before the great deluge).[27] Solomon was given skills to drive away demons and incantations to cure and alleviate diseases.[28] These skills came from a ring Michael gave Solomon at the commandment of God.[29]

Figure 20: Solomon Ring

Was Solomon a type of magician with a long beard, who held a staff, and who wore a warlock's hat, like the images seen in movies today? No.

The Egyptian word *heka* translates to 'magic.' Transliterated to Arabic, this word is *hekma*, meaning 'wisdom,' showing that magic is not as it seems. The magic arts, mystical or ancient mysticism, is rooted in the understanding of unity between all things, especially the spiritual and earthly.[30] These were the wisdom teachings found throughout the ancient world. Yahweh grants Solomon wisdom:

"Behold, I give you a wise and discerning mind so that none like you has been before you and none like you shall arise after you." – 1 Kings 3:12 ESV

His wisdom is seen:

"And people of all nations came to hear the wisdom of Solomon, and from all the kings of the earth, who had heard of this wisdom." – 1 Kings 4:34 ESV

The *Apocrypha* book *Wisdom of Solomon,* also known as *Wisdom,* shows this wisdom:

"For this cause, I prayed, and understanding was given to me. I asked, and a spirit of wisdom came to me... For [god] himself gave me an unerring knowledge of the things that are, to know the structure of the universe and the operation of the elements; the beginning, end, and middle of times; the alternations of the solstices and the changes of seasons; the circuits of years and the positions of stars; the natures of living creatures and the raging of wild beasts; The violence of winds and the thoughts of men; the diversities of plants and the virtues of roots. All things that are either secret or manifest I learned, for wisdom, that is the architect of all things, taught me." – Wisdom 7:7, 17-22.

This passage from *Wisdom of Solomon* echoes the wisdom or advanced knowledge in the *Book of Enoch*. This knowledge was known and taught in ancient mystery schools in Egypt. It was written in hieroglyphics before the flood to ensure that advanced knowledge from Atlantis and Lemuria would be remembered. Hermetic wisdom taught unity between the earthly and spiritual realms.[31] Hermeticism and Christianity existed harmoniously alongside one another in the early development of Christianity. It was not until the teachings of St. Augustine that hermeticism and paganism began to splinter. St. Augustine taught the incompatibility of the two and even referred to hermetic doctrines as demonology, pagan idolatry, incantations, and magic.[32]

H. P. Blavatsky explains this magic as it has come to be known in her book *The Secret Doctrine*. She notes that the entire biblical narrative is a historical account of the struggle between white magic (the mystery teachings) and black magic (the Levites, clergy, or indoctrination).[33] Ancient mystery teachings are imbedded into biblical teachings, as chapter nineteen will touch on. This new information shows that Solomon wore the masonic ring or the "Star of David" (pentagram/hexagram),[f] performed magic, and cast out demons. This is also known as the figure of Solomon, who was part of the mystery school and wore a ring showing his affiliation. The knowledge he received gave him the ability to understand advanced knowledge (magic).

Madden Jones speculates on Solomon's birth. In *Wisdom*, Solomon wrote:

"I molded into flesh in the time of ten months in my mother's womb, being compacted in blood from the seed of man and pleasure that came with sleep." – Wisdom 7:7

The ten months in his mother's womb is odd because the normal gestational period is nine months. The story from the *Atra-Hasis* gives the creation of man during the cloning phase of ten months in the womb of Ninhursag. 1 Chronicles 28:5 David quotes Yahweh: "And all of my sons (for the Lord has given me many sons) he has chosen Solomon my son to sit on the throne of the kingdom of the Lord over Israel." Succession usually goes to the firstborn son.

Another clue about Solomon's origin is in the *King James Version* of Psalm 45:2: "You [Solomon] are fairer than the children of men." The

---

[f] The five points of the star represented Sirius (Pike. *Morals & Dogma*, p. 17.).

skin tone of the region was negroid. Yet Solomon is stated to have a fairer complexion. It also references "children of men," pointing the reader back to the time of the nephilim and Watchers. As Madden Jones speculates, this means that extraterrestrial beings could have engineered Solomon as a hybrid, brought forth as a second child through Bathsheba, and chosen to rule over Israel to continue Yahweh's genetic lineage.[34]

The story of King Solomon is just one part of the continuation of mystery teachings and extraterrestrial contact that is vastly overlooked in the Deuteronomic cult's explanation. During the end of the 6th century, the priests stripped the meaning behind Solomon in Chronicles 1 and 2 to focus on legitimizing the new Yahweh cult.[35]

History paints Solomon as a wise and great king. Whether his story is factual or symbolic, the character of Solomon, King of Israel, gives a blueprint for moral and symbolic meaning. Solomon's Temple is filled with as many questions as the character of Solomon has. While Osman theorizes that Solomon and his Temple are an intentional misinterpretation of Amenhotep III's life and temple complex, others note that the design of Solomon's Temple was based on the Luxor Temple belonging to Amenhotep III.[36]

Solomon's Temple is Not What You Think

Mystery, symbolism, and millennia of unanswered questions surround the stories of Solomon's Temple. Was it real? Why has it not been rebuilt? What secrets lie beneath the surface of the stories across time? The temple, temple complex, tabernacle, and Ark are the only places in the Old Testament that detail specific dimensions to follow. Much like the importance of genealogy, when extra detail is given, it's time to pay attention and pull the string of curiosity because there is more information to unravel.

The biblical editors of 1 Kings 7:13-14 and 2 Chronicles 2:13 relay information about the architect or master craftsman known as Hiram of Tyre. This account is stated:

"And King Solomon sent and brought Hiram from Tyre. He was the son of a widow of the tribe of Naphtali, and his father was a man of Tyre, a worker in bronze. And he was full of wisdom, understanding, and skill for making any work in bronze. He came to King Solomon and did all his work." – 1 Kings 7:13-14 ESV

. . .

In 2 Chronicles, the name of this man is given as Huram-Abi.

Continuing the comparison between Solomon and Amenhotep III from the last chapter, Hiram, King of Tyre, sent Hiram Abiff, the craftsman, to assist Solomon in crafting the temple. Amenhotep, son of Habu, was the chief architect of Amenhotep III, who supervised the construction of Luxor, Karnak, Thebes, and the royal palace at Malqata. He was awarded the rank of chief scribe, was a steward of Queen Sitamun (the pharaoh's daughter and half-sister wife of Amenhotep III), and was in charge of recruiting.[1] Chapter eighteen explains the correlations between these two figures and their relation to another great ancient architect from Egypt, Thoth.

To adequately explain the various theories behind the temple and complex, each part is represented separately, inspected, dissected, and then placed back together with a greater understanding of the mystery of this building. The temple has three main parts: the Temple, the Holy of Holies (inner house), and the structures outside the temple.

THE TEMPLE

The dimensions and general description are given in 1 Kings 6:2-10: 60 cubits long (90 feet), 20 cubits wide (30 feet), and 30 cubits high (45 feet). A vestibule (porch) on the temple's east side formed the entranceway, was 10 cubits wide (15 feet), and extended along the width of the temple (30 feet). There were three stories of chambers along the sides of the temple. The first story was 5 cubits wide (7.5 feet), the second story was 6 cubits wide (9 feet), and the third was 7 cubits wide (12.5 feet). Each room was 7.5 feet wide, and unlike modern American architecture, rooms became larger the farther they were from the ground, creating a dome-like effect.

The temple was made of cedar and cypress and secured outside the house with offset beams so that the supporting beams could not be inserted into the temple's walls. Stones were quarried and brought to the temple for placement. No tools were used to create the temple (1Kings 6:1-10). Josephus remarks that the stone outside the temple was polished and smooth,[2] and Adam Clarke notes that the Vulgate translation stated that the temple was made of marble with the highest quality gold.[3] The inside of the temple was covered in gold and 2 Chronicles 3:5-7 notes that palms, chains, and cherubim were carved into the temple walls.

Little information is given about the temple's general structure besides the recessed framed windows. There could be many reasons for this, yet enough information is provided to begin unpacking the temple. Osman correlates the Temple of Solomon to the Luxor Temple of Amenhotep III. The Karnak inscriptions give insight into the various temples at Luxor:

"I build for thee thy house of millions of years in the [---] of Amon-Re, lord of Thebes (named): Khammat august in electrum, a resting place for my father at all his feasts. It is finished with fine white sandstone; it is wrought with gold throughout; its floor is adorned with silver, all its portals are of gold."[4]

Ancient history shows that electrum was a term used to denote amber. It was called gold because of its sunny brilliance. The *London Encyclopaedia* notes that electrum was used because it helped to detect poisons. *Electrum* is the Latin version of the Greek word *electron*. Biglino notes that the properties of the electron were known in ancient Greece.[5] Osman also repeatedly notes that the Second Commandment forbade idols to be made or images to be carved.[6] This causes an idea to stir in the back of the mind: if they were forbidden to make images, then what were the images of, or rather, what was the significance of reading about the palm trees, chains, or cherubim?

Figure 21: Layout of Solomon's Temple
Credit: MacKey, Albert. *Encyclopedia of Freemasonry and Kindred Sciences Vol II*, p.768.

Hall discusses five types of trees that held significance in the ancient world: oak, pine, ash, cypress, and palm.[7] Mason tradition notes that the palm tree, specifically the dates produced by it, is associated with beliefs all over the world.[a,8] If the Israelites were not allowed to make images, and the Old Testament was describing the Luxor Temple, could the cherubim have been a description of hieroglyphs?

THE HOLY OF HOLIES

Also known as the inner house, the Holy of Holies contained a golden altar, a golden table for the bread of the presence, five lampstands facing north and five lampstands facing south, lamps, flowers, gold tongs, cups, snuffers, basins, incense dishes, five pan sockets of gold for doors, and two large cherubim. The two cherubim were made of olivewood, each 10 cubits high (15 feet), with a total wingspan of 15 feet wide. They were covered in gold along with the walls and floors of the innermost part of the house. Along the walls and doors were engraved cherubim, open flowers, and palm trees.

---

[a] When a phoenix is seated on a palm tree, it signifies resurrection and eternal life.

Holy of
Holies

Holy
Place

Porch

Figure 21: Layout of
Solomon's Temple
Credit: MacKey, Albert.
*Encyclopedia of
Freemasonry and Kindred
Sciences Vol II*, p.768.

Ann Madden Jones gives her opinion about the various items located within the Holy of Holies. She notes that the oil from the lampstands created a type of insulation on the golden walls. The Ark of the Covenant was set on the golden altar. Remember from chapter thirteen that the Ark was a type of holographic communications device and that it emitted a type of radiation deadly to all who came near it. She noted that this part of the temple was a highly charged environment. The burnt offerings offered a type of blanket of positive ions to counteract the negative ions from the Ark. The lampstands she refers to as a type of stereo microphone to amplify the Ark's message.[9]

The Cherubim

Just as the cherubim on the ark, the cherubim in the Holy of Holies are not what the church has taught. The angelic picture of the inside of this cubed room is that it is covered in gold, with two large angel-like beings facing towards the Ark of the Covenant with their arms and wings stretched out. The description given in 1 Kings reads as though someone was present during the construction and description of the temple. The type of specifications given do not paint a picture of this angelic inner temple, but something completely different.

The biggest problem that leads people to believe that the cherubim are angelic beings is the word used for 'wing,' *kanaf*, as Biglino notes in the original Hebrew. The wing is taken as a literal wing of a bird, yet *kanaf* refers more to a building's wing.[10] There is also no description of the body of the cherubim, including the width of the body. Using the power of observational detail given by the biblical editors, it is safe to say that the body of the cherubim either does not exist or is simply something that separates the wings. It is time to discard the image of these beings that the church has given and create something new. Using one of the definitions Biglino gives for cherubim—a type of capacitor for a mechanical structure—and interpreting the word 'wing' as 'covering,' we can rewrite the description as:[11]

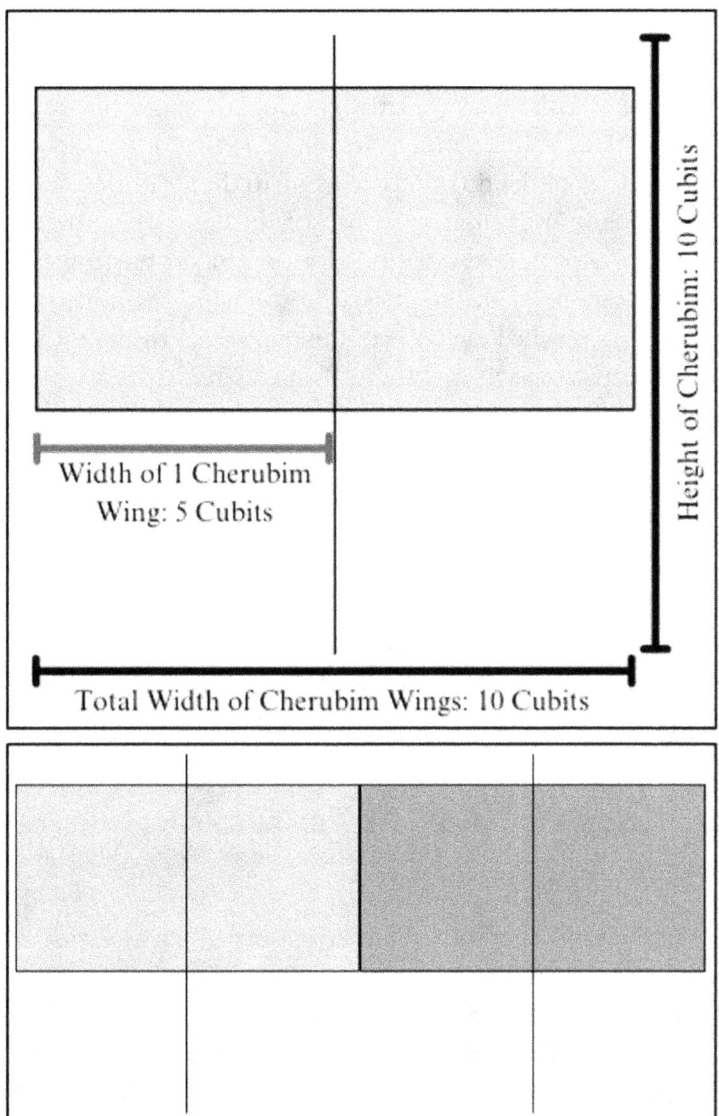

Figure 22: (Top) Artist interpretation of Single Cherubim with dimensions.

(Below) Artist interpretation of Two Cherubim touching wing to wing in the Holy of Holies

"In the inner sanctuary, he made two capacitors of olivewood, each ten cubits high. Five cubits was the length of one covering of the capacitor, and five cubits was the length of the other covering of the capacitor; it was ten cubits from the tip of one covering to the tip of the other... He put the capacitor in the innermost part of the house. The coverings of the capacitors were spread out so that a covering of one touched one wall, and a covering of the other capacitor touched the other wall; the other coverings touched each other in the middle of the house. And he overlaid the capacitors with gold." – 1 Kings 6:23, 27-28

This new description gives a picture vastly different from that of an angel with its wings stretched out. The visual picture in Figure 22 shows a type of panel used to block or cover something. Madden Jones describes the cherubim as plates that reflect and block charged oscillating electrons. Adding to her description of the Ark and the rest of the Holy of Holies, the positively charged ions from the lampstands are blocked by the negatively charged surface of the cherubim.[12]

Structures Outside the Temple

The Pillars

The mysterious pillars of the temple have spawned questions for millennia. At first glance, they appear to be just pillars, yet many people suggest otherwise. The description of these questionable objects is found in 1 Kings 7:15. They are pillars of bronze and are 18 cubits tall (27 feet), 12 cubits in circumference (12 feet), hollow, and four-finger width thick. An important discrepancy is in 2 Chronicles 3:15, which notes that the pillars were 35 cubits high (52.7 feet).

Adam Clarke points out that the pillars of the Temple were strictly symbolic and not used as structural support for any part of the building.[13] Osman notes that it was customary among ancient Middle Eastern cultures to name sacred objects. He postulates that these pillars were the Colossi of Memnon or the two Amenhotep III statues that stood before his temple.[14] Osman does not give an in-depth perspective of the pillars but only notes their naming and possible Egyptian correlation. Albert Pike gives the symbolism of the two pillars as a visual representation of duality: good and bad or the light and dark present in this reality.[15] A comprehensive breakdown is needed to fully understand what these pillars could have been.

The Capitals

The biblical description of the capitals that were formed on the top of the pillars is given in two places as follows:

"Then he made two capitals of cast bronze, to set on the tops of the pillars. The height of one capital was five cubits, and the height of the other capital was five cubits. He made a lattice network, with wreaths of chainwork, for the capitals which were on top of the pillars: seven chains for one capital and seven for the other capital. So he made the pillars and two rows of pomegranates above the network all around to cover the capitals that were on top, and thus he did for the other capital. The capitals which were on top of the pillars in the hall were in the shape of lilies, four cubits. The capitals on the two pillars also had pomegranates above, by the convex surface which was next to the network, and there were two hundred such pomegranates in rows on each of the capitals all around. The tops of the pillars were in the shape of lilies. So the work of the pillars was finished." – 1 Kings 7:16-20, 22 NKJV

"The two pillars, the bowls and the two capitals on the top of the pillars, and the two lattice-works to cover the two bowls of the capitals that were on the top of the pillars, and the four hundred pomegranates for the two lattice-works (two rows of pomegranates for each lattice-work to cover the two bowls of the capitals that were upon the top of the pillars)." – 2 Chronicles 4:12-13, Tree of Life Version

The Masonic tradition found in *A New Encyclopedia of Freemasonry* gives a symbolic perspective to the pillars. Two interpretations are given for the chapiter: a globe or spherical body,[b] and two crowns joined together.[c] The Masonic symbolism behind the pillars represents the sustaining power of God, while the globe-like chapiter symbolizes the body of the earth. Regardless of the interpretation of the chapiter, only the top half was decorated.[16] Madden Jones notes that the chapiter was that of a bowl-type structure, fusing both Hebrew words.[17]

Lily work

While the description in 1 Kings 6:19 references lilies or lily-work, the Masonic tradition states that the top of the capital consisted of a border of lilies that faced downward. It also clearly states that the pillars in Solomon's Temple were copied from those in Egyptian temples. The lilies are then a type of Egyptian sacred flower known as the lotus of the Nile, representing God's power to secure safety in the world.[18] Pike echoes this interpretation and notes that the capitals were constructed in the shape of the lotus seed vessel, or Egyptian lily.[19] The way that the description is written is:

"Now the capitals that were on the tops of the pillars in the vestibule were of lily-work four cubits." – 1 Kings 6:19 ESV

This description, along with the *New King James Version*, creates the image that the sides of the bowl or chapiter curved out like a lily 4 cubits wide (6 feet).

The Lattice Network and Wreaths of Chainwork

---

[b] Interpretation by Rabbi Solomon uses the Hebrew word *pommel*, meaning 'globe.' Pommel is used in 2 Chronicles 4:12 NIV, NKJV

[c] Interpretation by Rabbi Gershom uses the Hebrew *koteret* with the root *keter*, meaning 'crown' or 'diadem.'

Not every translation gives the detail of both the lattice network and a wreath of chainwork; most blend them together. Waite suggests that the lattice of checker-work is simply a network of fabric similar to that of a net. He uses the Hebrew word *gedil* (twisted threads, tassels) and speculates that it was a type of fringe material or tassel that was attached to the capital and hung down, similar to that of the priest's robes.[20] But the biblical description notes that everything that was on the pillar was cast in bronze and could not have been a physical fabric that was laid over the capital or was hung off the side.

Madden Jones incorporates the description into her interpretation and remarks that instead of being fabric it was a type of metal mesh known as a net of checkerwork or a wreath of chainwork. She suggests that the metal mesh covered the entire bowl-shaped capital. This acted as a type of network and, when struck by incoming microwaves, produced a movement of electrons. She uses the biblical description and notes that one end of the checkerwork could have been tucked into the inside of the cylinder pillar through a type of drain hole. The inside of the pillar could have been electrically neutral and assisted the negative electrons to flow from the chainwork, down the pillar, and towards the temple, causing a net negative charge.[21]

The Pomegranates

Manly P. Hall notes that the pomegranate on the capitals was used as a symbol. The pomegranate fruit was a mystic fruit in ancient times and associated with the Eleusinian rites.[d] When the fruit was eaten, it symbolized man becoming mortal or being deprived of immortality. The flower of the pomegranate was often seen with Greek gods and goddesses and symbolizes them giving life.[22]

Madden Jones once again takes a literal approach and theorizes that the pomegranates are instead a reference to a shaped object. She uses the

---

[d] Britannica notes these rites were the secret religious rites of ancient Greek mystery schools.

shape of a small brass sphere that covers the entire chainwork. These metal spheres were used as microwave-focusing devices for the metal mesh covering.[23]

The Name: Jachin and Boaz

There are so many explanations behind the names of the pillars. As Figure 23 shows, there is more than the common story tells. The only concrete detail about the pillars is:

"[Hiram-Abiff] set up the pillar on the south and called its name Jachin, and he set up the pillar on the north and called its name Boaz." – 1 Kings 7:21 ESV

The meaning behind the names is only given in commentary within the Bible or outside commentary. Jachin is composed of two Hebrew root words and breaks down as *Jah* (Jehovah) and *achin* (to establish), signifying 'God will establish.' Boaz is also composed of two Hebrew root words *b* (in) and *oaz* (strength), signifying 'in strength.'[24] Together, the names could mean 'God's strength will be established.'

Waite references the Kabalistic treatise called *Gates of Light* and notes that those who understand the meaning behind the name Jachin and Boaz will understand the *Neshamoth* (mind), the *Ruach* (spirits), the *Nephasoth* (souls), *El-chai* (living god), and *Adonai* (Yahweh). This means that to understand the pillar's significance is to understand the aspects of human life and its connected divinity to the formless.[25] This description gives insight into the possible mystical aspect of the Temple of Solomon.

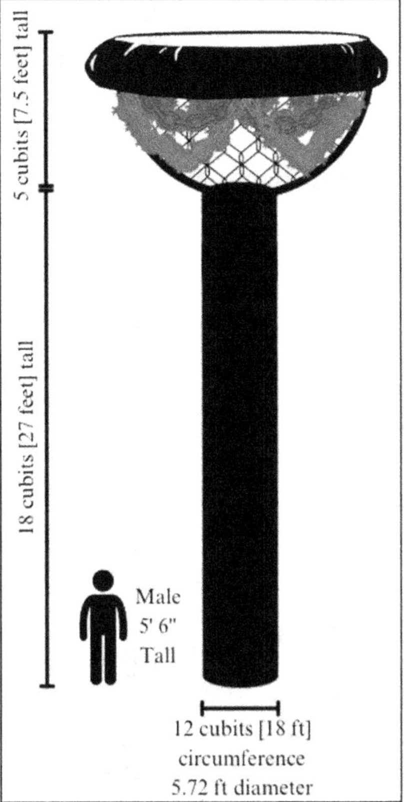

Figure 23: Artist interpretation of Pillar

Provided: Size perspective with an average human height of a 5'6" male

Pike offers another perspective from within Freemasonry. He notes the probable pronunciation of Jachin is *Ya-kayan*, meaning 'strengthening,' 'firm,' 'stable,' or 'upright.' While Boaz or Baaz means 'strong,' 'strength,' 'power,' 'might,' 'refuge,' 'source of strength,' or 'fort.' Concluding the meaning behind the names: Jachin is an active and animating energy or force, and Boaz is a passive type of stability or permanence.[26]

Madden Jones gives a very different perspective from Waite but is closer to that of Pike. When objects are named in the Bible, it tells their importance. The fact that each of the metal columns has a name and the intricate detail of their creation points to something more than just ornamental and symbolic. These were actual structures and were vital in the function of the communications complex known as Solomon's Temple. Instead of Jachin and Boaz symbolizing God establishing his strength, "it shall establish" refers to contact, and "in it is strength" refers to electrical power, noting that in the pillars, contact is made through electricity.[27]

Ten Basins of Bronze

1 Kings 7:27-39 describes each basin and its stand in detail, just as all the other details of the courtyard and Temple:

- Each stand was 4 cubits long, 4 cubits wide, and 3 cubits high (6 feet wide, 6 feet long, and 4'6" high).[e]
- Each stand had four panels set in frames:
    - Each panel featured a cherubim, lion, and oxen, with wreaths of beveled work[f] above and below.[g]
        - Chapter thirteen helps to solve the problem of what these wreaths were. It recalls that the Hebrew origins of the phrase "beaten work" that described the cherubim wings of the Ark were to be molded and hammered or knotted into a coil-like formation.[28] This same conclusion should be used with the "beveled work," as noted above.
- Each stand stood on four bronze wheels that were a 1.5 cubits tall (27 inches)
    - The bronze wheels resembled chariot wheels.

---

[e] Josephus notes a variation in the dimensions: 5 cubits long x 4 cubits wide x 5 cubits high (*Antiquities of the Jews*, 8.81).

[f] Other translations state: hammered work (NIV), wreath decorations (NLT), thin work (KJV), wreaths of plaited work (NKJV), bands of brass hanging down (Douay-Rheims Bible), and spiral figures (Good News Translation).

[g] Josephus notes a lion, bull, and eagle with no mention of a cherubim (*Antiquities of the Jews*, 8.82).

- Each stand had bronze axles attached to the stand and not the wheels.
- Each stand had four corners to support a basin, cast with wreaths on each side.
- The basin sat on top of a crown that was 1 cubit and projected upward with a pedestal that was a 1.5 cubits deep (2'3").

- Each basin was placed on an attached round band 0.5 cubit high (9 inches) and was cast into the stand.
- Each bronze basin was 4 cubits tall (6 feet) and held 40 baths.
- Five basins were placed on the south side of the temple and 5 on the north.[h]

[h] ESV translation.

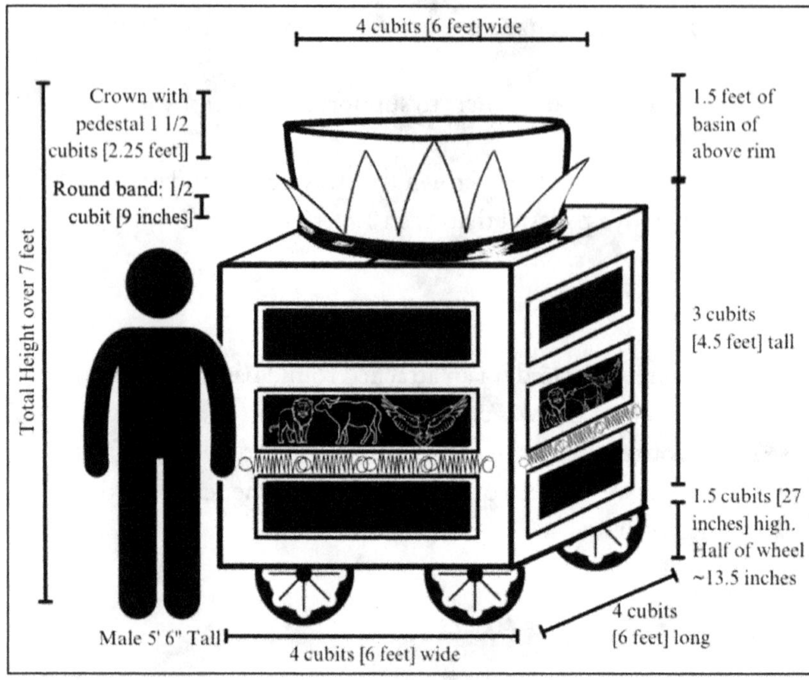

Figure 24: Artist's Interpretation of single bronze basin

Provided: Size perspective with an average human height of a 5'6" male

Figure 24 is not an exact scale, yet it gives an idea of what the individual lavers could have looked like. Madden Jones gives various dimensions and ultimately concludes that each laver was over 7 feet high, noting that the individual lavers or wash basins were not used for any type of cleaning at all. While 1 Kings 7:39 notes that each basin or bowl could hold 40 baths, this is a type of measurement, not necessarily a quantity. Today, water is measured in weight (gallons, pounds, or kilograms), and forty baths is approximately 340 gallons, weighing over 2,700 pounds. This incredible weight was held up by a four-inch-thick solid metal bowl with a diameter of 6 feet.[29] The commonly held belief is that these lavers were used for washing the sacrificial meat. If this were the case, there would be no need to have each of the ten lavers be portable, and instructions would have been given for ladder construction with rituals to wash the meat in the laver. The absence of this evidence, paired with the incredible size and weight, leads one to believe that there must have been another use.

What if, instead of being used for washing, as previously believed, these lavers were used as various portable antennae that could be moved around the courtyard to ensure the best signal to receive information from an orbiting ship?[30]

The Molten Sea

The last piece of the Temple puzzle is the sea of cast metal, also called a molten or brazen sea. A molten or brazen sea is a large vessel to hold water.[i] The description is:

"Ten cubits (fifteen feet) from brim to brim, five cubits high (seven feet six inches), and a line of thirty cubits measured its circumference (forty-five feet)." – 1 Kings 7:23 ESV

Just like the chapiter, the brim of the metal sea flared out like that of a lily and a handbreadth thick, or four inches (1 Kings 7:26). It is easy to gloss over, but this is not a perfect circle or bowl. The diameter of a circle with a circumference of 45 feet is 14 feet, not 15; the circumference of a circle with a diameter of 15 feet is 47, not 45. This then causes the shape of the sea of bronze to create a parabola, like a modern-day rooftop satellite.

Next, depending on the translation, there are gourds or knobs spaced out every ten cubits under the brim of the metal sea, all cast together as one piece (1 Kings 7:24). In other words, there are two rows of objects that resemble gourds, made of bronze, and set onto the sea.

The metal sea was placed on top of 12 oxen, with the rear of the oxen facing inward. Three oxen faced north, three faced east, three faced

---

i www.biblegateway.com/resources/encyclopedia-of-the-bible/Molten-Sea

south, and three faced west. The 12 oxen are said to represent the 12 symbols of the zodiac looking towards the four cardinal points.[31]

This large metal sea is to hold "two thousand baths" (1 Kings 7:26). Like the portable lavers, it is used as a measurement of reference and should not be taken literally. If it were literal, the sea would have held 17,000 gallons of water, weighing a staggering 141,610 lbs. Just as with the portable lavers, no instructions were given for a ladder, rituals, or how to clean the sea.

Given the information from 1 Kings 7, the only logical explanation is that there is more than meets the eye, and it has nothing to do with water. What if, instead, the molten sea was a type of stationary transmitting satellite?

RECREATING THE TEMPLE COMPLEX

Madden Jones creates this argument in *The Yahweh Encounters*. The Ark causes the Holy of Holies to become a highly electrically charged environment. The oil from the lampstands provided insulation as the burned oil coated the gold-covered walls and floor. Burnt offerings were offered in the courtyard just outside of the Holy of Holies, and the smoke provided electrically positive ions to counteract the negative ions given off by the Ark. The two cherubim acted as ionic speakers, and the candlesticks became a microphone to speak into. This was all powered by the pillars or an ancient version of a Van de Graaff generator. When struck by arriving microwaves, electrical ions traveled down the electrically neutral pillars, increasing the negative charge and providing power to the entire temple complex. While there is not much information about the windows or side rooms within the temple, Madden Jones speculates that the light received from the windows, coupled with the possible empty space of the rooms, acted as a type of frequency dial. Similar to a tuning dial, if a higher frequency was needed, light or space could have been adjusted.

Figure 25: Above: Brazen Sea. Artist Interpretation from Jewish Encyclopedia, 1906.

Credit: Public Domain, Wiki Commons.

Right: CSIRO Earth Receiving Station Satellite Dish Droughty Point Hobart Tasmania.

Credit: Bruce Miller CC BY-SA 4.0, Wiki Commons

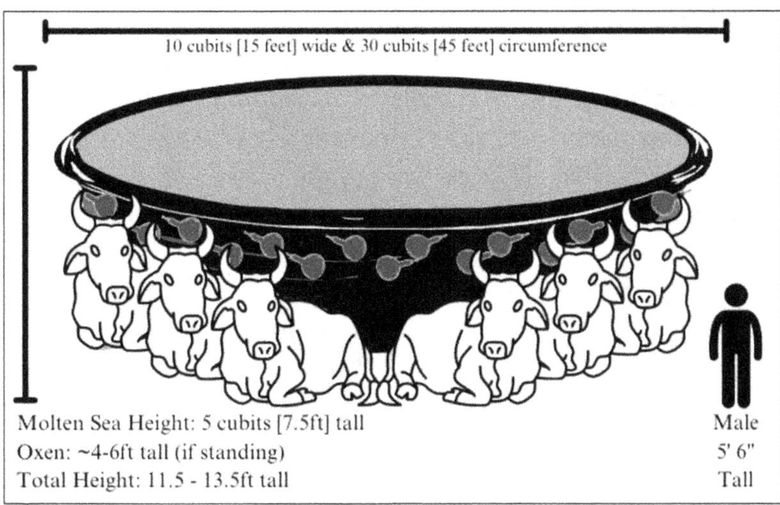

Figure 26 gives a visual of the entire temple complex. The movable lavers acted like portable receivers, allowing them to move around the courtyard to find the strongest signal—just like how people today walk around to find better cell service. The stationary metal sea acted as a transmitting satellite to send messages.[32]

This entire new picture of Solomon's Temple makes one think of it as an upgrade from the tabernacle and Ark from Moses' time. Instead of just an audible holographic communication system, the temple was a state-of-the-art, stationary, visual, and audible holographic communications center. This new communications center allowed direct communication with a mothership that hovered just above the atmosphere, as chapter nineteen will show. A glimpse is given into the temple's true purpose:

"Now my eyes will be open and my ears attentive to the prayer that is made in this place. For now, I have chosen and consecrated this house [Solomon's Temple] that my name [Yahweh] may be there forever." – 2 Chronicles 7:15-16 ESV

This new viewpoint, along with all the information to this point in the book, shows that advanced technology existed in mankind's past. Josephus makes an interesting comment in *The Antiquities of the Jews* and remarks that Hiram made mechanical works for the temple based on the plans Solomon gave him.[33] The information has been suppressed, but the truth is beginning to peek through the cracks and will ultimately unravel the carefully constructed narrative of the Christian Church.

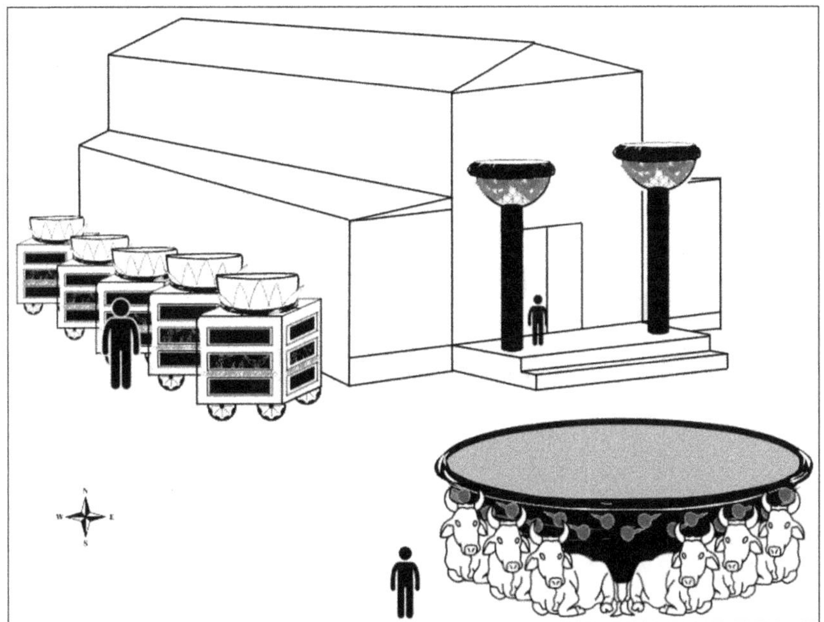

Figure 26: Artist Interpretation of Solomon's Temple Complex

Provided: Size perspective given with three average-height males

The Strange Enigma of Hiram

As previous chapters have shown, there is a reason for everything written in the Bible. The purpose, however, is unclear until one digs deeper into the characters introduced and the events surrounding them. Most stories are read without prior knowledge or outside interpretations. Each new character reveals a piece of the larger story. A character may seem meaningless until the end when the reader finally comes to the climax and finds out that the mysterious character who took up only three lines was behind everything. This is seen in *The Wizard of Oz*, where the Wizard is the one who is controlling everything but is only in the last few minutes of the entire movie.

The Bible is not read like a story; it is meticulously taken apart and studied in pieces without understanding the correlation between stories. To add to the confusion, most people have a general understanding of the stories in the Bible without ever having read a single page. Solomon's Temple is discussed so often with many various interpretations and visual depictions that it becomes easy to miss all the intricate details that are explicitly stated in the text. Biblical translators throughout the millennia have also altered the meaning of specific words to the point that the initial story is unrecognizable today. The story of Hiram is much the same.

The account of Hiram, the architect of the Temple of Solomon, comes from the Bible, which gives very little information. However, Egyptian and Masonic sources give insight into a minor character and reveal him to be one of the largest characters on the world stage of myths and legends.

THE BIBLICAL ACCOUNT

Appearing only in 1 Kings 7 and 2 Chronicles 3, the Bible gives little information about Hiram other than that he worked on the temple. The first introduction is that he:

"Was the son of a widow of the tribe of Naphtali, and his father was a man of Tyre, a worker in bronze. And he was full of wisdom, understanding, and skill for making any work in bronze." – 1 Kings 7:14 ESV

The second time his name appears, the biblical editors noted that he was a "skilled man, who has understanding, Huram-Abi, the son of a woman of the daughters of Dan, and his father was a man of Tyre. He is trained to work in gold, silver, bronze, iron, stone, and wood, and in purple, blue, and crimson fabrics, and fine linen, and do all sorts of engraving and execute any design that may be assigned to him, with your craftsmen" (2 Chronicles 2:13-14). Neither of these descriptions gives much insight into Hiram or why his assistance with the temple is so important.

Josephus gives slightly more information and notes that his mother was from the tribe of Naphtali, and his father was from Ur, of the Israelites. He was skilled in all types of work but was *chiefly* skilled with gold, silver, and bronze.[1] As noted in the last chapter, he was also gifted in mechanics. What an odd piece of information to give about a man in what was thought to be "primitive" times, or rather, before machines were invented.

The Egyptian Account

If the stories of the biblical patriarchs match that of the 18th Dynasty pharaohs, Amenhotep III aligns with Solomon. Amenhotep, son of Habu, would meet the qualifications of Hiram the architect. The stories of Amenhotep are found at the Dayr al-Madīnah. He was born circa 1450 BCE to Habu and Itit[a] during the reign of Thutmose III. During his life, he was known for his intelligence and moral integrity. His first public position was Inferior Scribe, and he was promoted to Superior Royal Scribe. At the age of 50, Amenhotep III promoted him to Minister of Public Works. While in this position, Amenhotep, son of Habu, led the construction of the Colossi Memnon in Thebes, the temple at Luxor, and the Soleb temple in Nubia. He organized the king's jubilee celebration—taking over the duties normally ascribed to the king's eldest son, Akhenaten—and was the steward to Sitamun, the king's daughter. He served under Amenhotep III for approximately 30 years until his death.

Masonic Tradition

The king and the architect are both named Hiram and come from Tyre. Because of this, each character needs to be understood separately.

Hiram, King of Tyre

The name Hiram means noble born in Hebrew, but it is an improper translation based on the Hebrew characters and should be rendered as Khurum.[b,2] A possible reason for this mispronunciation is found in the Arabic root Ḥ-R-M,[3] meaning 'of which will be revealed.' This king's friendship with Solomon is seen in Biblical sources such as Josephus, who explains the origin of their friendship. When Abibaal, King of Tyre, died, Hiram became king, and at the age of 53 he cut down mate-

---

a *Ancient Records of Egypt Vol II* notes his mother's name was Yatu.

b Pike notes Khurum breaks down to *khi* (living) and *harūm* (shall be raised or lifted up) with the full meaning of Khurum, 'raised up to life.'

rials to build the temple of Hercules and Astarte (Asherah). The temple of Hercules was completed in the month of Peritios.[4] The second day of Peritios (Barith) is December 25, and Josephus tells in Antiquities 8.146 the story that is found in the Annals of Tsūr. When the temple of Herakles was erected by Hiram (Khurum), marks the first festival of Melqart, celebrating the Sun during the winter solstice, when the rebirth of the sun god regains a new life by fire. A festival also seen in Rome and known as *Dies Natalis Solis Invicti* (Festival Day of the Invincible Sun).[5]

Hiram's success in building the Temple of Herakles may have been what prompted Solomon to seek out workmen from Tyre to assist with the temple's construction.[6] The Phoenicians of Tyre[c] and Sidon were known as the best mechanics in the world,[7] and Hiram, King of Tyre, was a Masonic Grand Master.[8] When Solomon asked Hiram, King of Tyre, to send someone to assist with the temple, Solomon understood that Hiram would have sent someone who was a master in their craft. This is exactly what Hiram, King of Tyre, did when he sent Hiram a Grand Master of Dionysiac Architect, Hiram Abiff.[9]

Hiram, the Architect

Unbeknownst to the non-Masonic initiate there is an incredible amount of symbolism with this character. While his father was from Tyre, his mother's origin is confusing due to the two descriptions given: "widow of the tribe of Naphtali" (1 Kings 7:14) and "woman of the daughters of Dan" (2 Chronicles 2:13). Mackey assists with this confusion and notes that she was from the tribe of Dan and married a man from the tribe of Naphtali who was possibly Hiram's birth father. She then remarried a man from Tyre who helped raise Hiram and thus became like a father to him.[10]

The Bible version also helps us understand what Abiff (Masonic tradi-

[c] Dionysiac fraternity of artifices was seated in Tyre (Mackey. *Encyclopedia of Freemasonry and Its Kindred Sciences Vol. I*, p. 330.).

tion uses 'Abiff' and rejects the English translation of 'Abi') means in 2 Chronicles 2:13. The *King James Version* reads, "of Huram *my father's,*" while the New International Version reads, "Huram-Abiff." Abiff is a compound noun of *Abi* (*father*) and *if* (*his*) so the translation of Abiff as 'his father' is accurate.[11] For the word *Abi* in Hebrew, the possessive pronoun precedes the noun, and 'of' is indicated by a dash (-), as in *Abi-Al*, meaning 'Father of Al.' The primitive noun in the Semitic *Ab* also means father, which is seen in Genesis 45:8 when Joseph refers to himself as *Ab l'pharaoh*, 'father to pharaoh.'[12]

Another—and more appropriate—use of the word *Abiff* is that of a title or designation that, when translated from Hebrew *ab,* means 'counselor,' 'wiseman,' 'master,' 'originator,' 'inventor,' 'master,' 'prophet,' or 'teacher.'[13] This new understanding shows that Hiram Abiff was a highly skilled man, or as noted earlier, a Grand Master of the Dionysiac Architect.

The Enigma of the Temple's Architect

Amenhotep, son of Habu

Egyptian and Greek sources of Amenhotep, son of Habu, show that he was known for his intelligence, wisdom, morals, and ethics during his lifetime, which he stated came from the teachings of Thoth. He was so highly regarded that the pharaoh allowed him to create self-portrait-type statues in various temples of Amun. These statues depicted Amenhotep, son of Habu, as an intermediary between people and God.[14] After his death, Amenhotep III created a mortuary temple and had a statue erected of him at the Karnak temple of Amon. The statue had an inscription noting his accomplishments in life and naming him the sole companion of the king and a prophet of Horus.[15] Along with these inscriptions, there was an assigned clergy to tend to this temple.

During the New Kingdom, until the end of the Third Intermediate and Later Periods of Egypt, a mortuary cult formed and declined. Yet, Amenhotep, son of Habu, remained independent and his importance increased. He became honored by Thebans and remembered for his wisdom in writing and literature. Statues from this time period show royalty viewed Amenhotep, son of Habu, as a good physician and a healer.[16]

Around 300 BCE, his mortuary temple was moved to Deir el-Bahari near Hatshepsut's temple. This new temple is the first documented place that shows Amenhotep, son of Habu, as a god. His father, Habu, was changed to the god Apis and was given the spiritual father, Thoth, the god of wisdom. His mother, Itit, was changed to Hathor, and his spiritual mother became Seshat, the goddess of writing. Along with these changes, it became Amun who made the memory of his architecture immortal, and he was sometimes referred to as a son of Amun.[17] Łajtar overtly notes that Amenhotep, son of Habu, was given the status of a saint in Egypt (one of only two people who obtained this status)[d] while it was Greek sources that called him a god.[18]

His name and connection with the gods had solidified in the ancient world. Josephus spoke about him during his discourse about a fictitious king of Egypt. The fictitious king's name was Amenophis, who communicated with the son of Papis by the same name, Amenophis. He was divine in wisdom, foresight, and fortune-telling, especially when it came to dealing with the Avaris before the birth of Osarsiph.[19] This "fictitious" story from Manetho, as chapter twelve noted, was about Amenhotep III, the Israelites, and Akhenaten.

Regardless of his status in either culture, he was regarded as wise in both. Inscriptions were found on Greek limestone pottery called *Polyaratos ostracon* from the 3rd century BCE. The inscriptions were nine statements, or ten sentences, that share ethical truths.[e] This ostracon may have been a propaganda tactic used by the priests of Amenhotep, son of Habu. Yet, it successfully presented him as a healer god and inventor of wisdom.[20] The most interesting detail on this

d Imhotep was the other mortal, non-royal who obtained the status of a saint.

e Breasted notes "nine fragmented sayings" while Łajtar notes it was titled "Commandments of Amenothes" and contained at least ten sentences that related ethical truths, or maxims that were popular in Greece then. He also notes that there was no correlation between the Egyptian wisdom literature and Greek maxims. The existence of the two is purely coincidental (Breasted. *Ancient Records of Egypt Vol II.*, p. 911; Łajtar, "Deir El-Bahari in the Hellenistic and Roman Periods", p. 26).

ostracon is that it contains sayings from what Wilcken calls the 'Proverbs of the Seven Wise Men.'[21] These seven wise men from the ancient world could only be a reference to the Seven Sages of Antiquity from chapter five, whose teachings were documented all over the world.

Hiram Abiff

As noted previously, the Arabic root of *Hiram* is HRM, which would have also been the correct Hebrew writing because the basic Hebrew script does not contain vowels. The Arabic word *Hîrm* means 'ox,' which is the symbol of the Sun in the Taurus constellation during the vernal equinox.[22] Chapter four told of the correlation between the Taurus constellation and the Pleiades. Osman notes that when *s* is added, the correlation is easily made, HRMS or another name for Hermes. Hermes Trismegistus ('thrice great') was the Greek god of wisdom and all-knowing sage of the ancient world.

Hermes was regarded as the king of all knowledge, a prophet, and a teacher who taught man the philosophy of universal unity between the spirit and the divine.[23] Hiram's title, Abiff, shows that he was a prophet, Wiseman, and teacher.

Hiram consists of three Hebrew consonants: *Cheth*, *Resh*, and *Mem*.

Letter	Pictogram[24]	Meaning[25]
Cheth ח		Braided rope, or possible DNA: Man with raised arms pointing towards something, to reveal: Tent wall, looking inside to reveal.
Resh ר		Head of a man Top, beginning, first (top of body), rule (chief's role) inheritance (decided by chief)
Mem מ		Pictogram is water meaning chaos, mighty, blood.

Table 6: Hebrew and Pictogram Root Word Association
Source: Benner. *The Ancient Hebrew Language and Alphabet*, p.119 and Biglino, *Gods of the Bible*, p. 67.

[24,25]

- *Cheth* symbolizes *Chamah,* the sun's light or the universal relationship between nature and the sun.
- *Resh* symbolizes the Biblical translation of *Ruach*, spirit, or the vehicle that collects the light and turns it into fire.
- *Mem* symbolizes *majim*, water. *Hiram* then, according to von Welling, represents Nature in one word.[26]

Dr. Sigismund Bacstrom, an initiate of the Brotherhood of the Rose Cross, found that the Chaldeans, Egyptians, and Hebrews have all drawn from one origin: *The Emerald Tablet of Hermes*. He found that Hiram's secret was that of the Philosopher's Stone.[27]

Hall remarks on the incredible symbolism of this seemingly minor character. He is the prototype of humanity and the archetype of man that Plato speaks of in *The Republic*. He also symbolizes the higher state that man aspires to in intellect, spiritual, and physical freedom. Lastly, Hiram Abiff is the resurrected Osiris or Christ who died at the hands of Jubela, Jubelo, and Jubelum (collectively called the Juwes), who represented the Libra constellation of scales held by Shamash, and whose names originate from the Chaldean god Bel.[28]

Whether the stories of Solomon and Hiram Abiff are factually true or not, one thing remains: Hiram's symbolic nature is greater than most can understand. Pike remarks that even if Hiram is an imaginary character, his character represents the ideal of what mankind can become and the possibility of this symbolism made real:

- Continuous progression toward the realization of one's true destiny,
- Intelligent,
- Noble,
- Incredible moral character.[29]

Osiris, Christ, and Hiram Abiff ultimately are resurrected into a different form and transcend into a higher state of consciousness. That is how this minor character, who has only two references, changes the entire story of the Bible. His name represents ancient knowledge from Atlantis, brought through the flood and carried throughout time in secret until the time is right or someone searches to find it.

Elijah, Ezekiel, Chariots, and Sky Armies

A narrative exists today that extraterrestrials are just now visiting us but history shows that they have been here all along. As previous chapters have shown, there are references to these visitations all throughout the Bible. The language and references have changed, but one fact remains the same: extraterrestrials have been visiting humanity since its inception. The stories of Elijah and Ezekiel have become over-symbolized, and what was once observational fact is now metaphorical teaching.

Elijah

Elijah was considered one of the greatest prophets of the Old Testament but could he have been more than just a prophet? Like Enoch, Elijah was taken into heaven:

"As they were walking along and talking together, suddenly a *chariot of fire* and horses of fire appeared and separated the two of them, and Elijah went up to heaven in a whirlwind." – 2 Kings 2:11 NIV (emphasis added)

. . .

Every translation of the Bible states this same event in the same verbiage.

This event was planned, and Elijah would meet Yahweh at the band of the Jordan River, as the preceding ten verses show. Elijah and Elisha traveled from Bethel to Jericho and down to the Jordan River. Each time they stopped, a group of prophets asked Elisha if Elijah was to be taken to heaven, and every time, Elisha told them yes and essentially to stop asking. When Yahweh came with his chariots of fire and took Elijah, Elisha called out:

"'Master! Master! Israel's chariot and horses!' and then asked, 'Where is Yahweh Elohim of Elijah?' The prophet's disciples of Jericho witnessed the event and saw 'Elijah's [*ruach*] rest on Elisha.' The disciples then bowed in front of Elisha. The disciples went to look for Elijah because 'Ruach Yahweh lifted [Elijah] up' and maybe dropped Elijah off in a surrounding hill or valley. They returned and found nothing." – 2 Kings 2:1-18 NOGV (summarized)

The passage answers what the chariots of fire are. Verse sixteen shows "the Ruach of Yahweh," also known as the UFO, is the chariot. Psalm 68:17 describes the chariots of fire from 2 Kings 2:1-18:

"The chariots of Elohim are twenty thousand in number, thousands upon thousands. Adonai is among them. The God of Sinai is in his holy place." – Psalm 68:17 NOGV

The Jerusalem Bible verifies that these are not the chariots of Solomon,[a] noting that they were not literal chariots or horses, but the heavenly

---

a 1 Kings 10:26 "And Solomon gathered together chariots and horsemen."

chariots seen by Elisha in 2 Kings 2:10 and 6:17—the mountain was covered with horses and chariots of fire—were a large group of UFOs. Elijah's amazing encounter with the chariots of fire was nothing more than an agreed-upon extraterrestrial abduction.

Ezekiel

There are many depictions of Ezekiel's chariots of fire or cherubim (the Bible version notes the difference in name), none of which come close to what is actually happening in the first chapter. Ezekiel's story begins with three verses of information that note his Jewish upbringing, good education, and familiarity with the cultures in the Middle East and Egypt:[1]

"The sky opened, and I saw visions from Elohim." – Ezekiel 1:1 NOGV

This suggests that something incredibly supernatural was happening when Ezekiel was at the Chebar River. However, this is completely false because the Hebrew word *maré* is mistranslated to 'visions,' giving it a type of ethereal phenomenon. Instead of etheric visions, Ezekiel had "*mareót* Elohim"—the plural form of *maré*—so a more accurate translation would be '*objective observations* of Elohim.'[2] One characteristic about Ezekiel is certain: he was a master at accurately describing what he observed, to the point that engineers today are able to use his descriptions and accurately understand what he was writing about.[3] Verse four continues with:

Hebrew
English
חַשְׁמַל
electricity

"I looked, and, behold, a *whirlwind* came out of the north, a great cloud with raging fire enfolding itself; and a brightness all around it, and radiating out of its midst like the color of *amber*, out of the midst of the fire." – Ezekiel 1:4 NKJV (emphasis added)

As noted previously, 'whirlwind,' 'wind,' 'spirit,' and 'breath' are all incorrect translations of *ruach*. The color of amber, Hebrew *chasmal*,[b] is used to describe the color within the fire. Table 9 shows that *chasmal* (חַשְׁמַל), when translated from Hebrew to English, means 'electricity.' Chapters thirteen and seventeen noted that the Greeks were well aware of the physical properties of the electron, and for some reason, even though the properly translated word is known, the concordance notes the origin is unknown.

The "great cloud and fire enfolding itself" represents the last phase of flight when the craft is reducing its speed.[4] This same visual can be seen when the SpaceX rocket gets close to the landing platform; a large cloud of smoke and fire forms around the rocket to decrease its speed. Verse 13 explains the color of amber when Ezekiel says, "Out of the fire went forth lightning" (1:13 KJV). Sparks of lightning are seen during electrical discharge during specific atmospheric pressure, high heat, and increased electrical density: all of the things occurring during the landing phase of a craft. The remainder of Ezekiel 1 describes three different parts of the craft:

1. The living creatures and wings.

2. The craft's lower shape.

3. The command capsule and occupant.

One thing that makes this description easy yet difficult to understand is our technology. Modern society has the technology to create parts of

---

[b] Strong's Hebrew Concordance no. 2830 notes uncertain derivation, possibly glowing metal.

this craft, but not all of the technology, especially the reactor core or central power unit for the entire craft.

The Living Creatures and Their Wings

Ezekiel describes exactly what he is seeing. While he does not have the technical knowledge to state what he is witnessing, his description is so precise that engineers took this information and created a technical explanation of it. Ezekiel's description notes that one item is described four times—the four living creatures.

The Firey Blaze and Wings

"Also from within it came the likeness of four living creatures. And this was their appearance: they had the likeness of a man. Each one had four faces, and each one had four wings [...] Their wings touched one another. The creatures did not turn when they went, but each one went straight forward [...] Thus were their faces. Their wings stretched upward; two wings of each one touched one another, and two covered their bodies. And each one went straight forward; they went wherever the spirit wanted to go, and they did not turn when they went. As for the likeness of the living creatures, their appearance was like burning coals of fire, like the appearance of torches going back and forth among the living creatures. The fire was bright, and out of the fire went lightning. And the living creatures ran back and forth, in appearance like a flash of lightning." – Ezekiel 1:5-6, 9, 11-14 NKJV

Ezekiel gives the likeness (Hebrew: דְּמוּת, 'character') for each of the four living creatures:

- Four faces
- Four wings
- The four wings touched each other

- Moves in all directions but does not physically turn
- Fire moves between all four
- Lightning appears between them
- Speed is faster than lightning

While the first three will be explained in the next sections, the main purpose is that the four

faces and four wings were present with each living creature. Each living creature was attached to something above, called the firmament, and moved with it faster than the speed of lightning. Each living creature contained fire, and electrical discharge came from them.

Blumrich offers a type of moveable column structure that contains all the features listed below. The column swings upward during the flight phase to increase aerodynamic capabilities. When the craft achieves a low landing velocity, the column rotates below the craft to perform its various functions, as explained below. The fire and lightning are explained as a type of control rocket that is seen in space flight today. The firing of these control rockets allows small adjustments of movement in space, just like the rudder on a ship or plane allows for small adjustments.[5]

Each column has four wings, two of which are straight up, and the other two cover the column. Once again, the word 'wing' is mistranslated from the Hebrew *kanaf* and would be better understood as 'extremity' or 'ends.' Ezekiel also notes that the *kanaf* were straightforward when the *ruach* moved.

This description is very confusing when trying to understand it from a metaphorical perspective. A technical perspective gives a very easy description of these straightforward extremities: propellors. When the craft was not using the propellors, they were folded, and when they were being used, they were straight out. Four propellers were attached to one column, creating a type of spherical helicopter. Four-column helicopters were attached to the main body for landing stability and short-distance flight. When the columns were not in use, they were flipped to the top of the craft to assist with the aerodynamic drag of descent and then flipped down when the speed decreased and the craft was preparing to land.[6]

The sound and position of the wings are described further down in the first chapter:

"When they went, I heard the noise of their wings, like the noise of many waters, like the voice of the Almighty, a tumult like the noise of an army; and when they stood still, they let down their wings." – Ezekiel 1:24

Ezekiel used the sound of water; other versions describe this sound as the noise of great waters or the sound of rushing water. This is an accurate audible representation of propellors as they hover in the air, and the sound associated with the *ruach* is consistently associated with the voice of God throughout the Old Testament. When the craft was standing still, the propellers returned to their original position during descent.

The Faces of Four Living Creatures

"As for the likeness of their faces, each had the face of a man; each of the four had the face of a lion on the right side, each of the four had the face of an ox on the left side, and each of the four had the face of an eagle." – Ezekiel 1:10 NKJV

Three of the four faces—the eagle, lion, and ox[c]—resemble the markings mentioned in chapter seventeen's description of the portable

c Other translations use 'bull' instead of 'ox.' The Hebrew [שׁוֹר] indicates a cattle's head and is gender neutral.

lavers. At this point, it becomes necessary to stop momentarily and notice the multiple references to the same three creatures:

- The Eagle
    - This could refer to the Aquila constellation. In Greek mythology, an eagle carries Zeus' thunderbolts.
    - The eagle is also the symbol of the Egyptian god Horus.[7]
    - The Sumerian eagle represents the flying gods from antiquity.[8]
- The Lion
    - The Sphinx's original design was as a lion before the face of the pharaoh was replaced over the lion's face.
    - The lion was the symbol of the god of Upper Egypt, Athom-Re.[7]
    - In Ugaritic texts, the lion was the symbol of Asherah.[9]
- The Ox (Bull)
    - Chapter eighteen remarked that Hîrm in Arabic means ox and symbolizes the Sun in Taurus during the vernal equinox. The Taurus constellation is the shape of a bull.
    - The Egyptian god symbolized by the bull is Apis.[7]
    - The bull was the symbol of the cult of Ba'al.[10]

Figure 27: Lunokhod 1

The four faces seen by Ezekiel could be:

1. The pareidolia[d] effect that can be seen in normal objects. Figure 27 shows this phenomenon with the Soviet Union's lunar rover Lunokhod 1, from 1970.
2. Painted images of the man, eagle, lion, and ox. If the Elohim are similar to humans, then it would make sense that some of their traditions of painting images on craft are mimicked today when pilots paint images on fighter jets.[11]
3. The face of a man could refer to the image of the Elohim, who look similar to humans.
4. Blumrich also notes that there were eyes on the outside of this column under the main craft. These "eyes" could be a type of visual monitoring similar to a camera lens today.[5]

[d] Pareidolia: the tendency to perceive a specific, often meaningful image in a random or ambiguous visual pattern (Merriam-Webster Dictionary).

. . .

The Hands of a Man

"The hands of a man were under their wings on their four sides; and each of the four had faces and wings." – Ezekiel 1:8 NKJV

Each of the living creatures or support column had hands. The hands that Ezekiel saw could have been mechanical hands attached to a mechanical arm. Figure 28 shows a mechanical arm used in space and an example of a robotic arm that looks very human-like. These could have been used in the same way that mechanical arms and hands are used today.

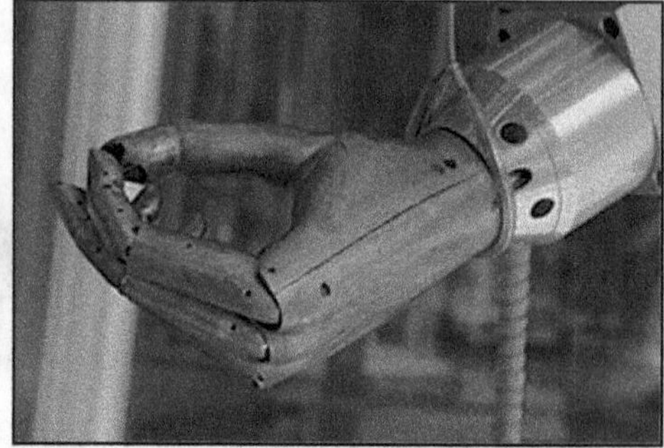

Figure 28: Above: European Robotic Arm Attached to Nauka

Credit: NASA, Public Domain, Wiki Commons

Below: MA-3 Robotic Manipulator Arm

The Straight Leg and Sole of a Calf's Foot

. . .

"Their legs were straight, and the soles of their feet were like the soles of calves' feet. They sparkled like the color of burnished bronze." – Ezekiel 1:7 NKJV

The straight leg Ezekiel describes turns into a straight, possibly telescopic, type of landing leg, with two for each column. The telescope function could be used as an ancient shock absorber to assist with landing. The sole of a calf's foot describes a type of disk at the end of the telescopic shock absorber. Nothing would keep the craft from sinking into the terrain if the small telescopic end landed on a soft surface. A disk-shaped end is needed to increase surface area and prevent sliding or sinking into the terrain. Blumrich also suggests that if it had an outward curve on the bottom, it would assist with traction and landing if excessive wind was present during landing. [5]

The Wheel Within a Wheel

"Now as I looked at the living creatures, behold, a wheel was on the earth beside each living creature with its four faces. The appearance of the wheels and their workings was like the color of beryl, and all four had the same likeness. The appearance of their workings was, as it were, a wheel in the middle of a wheel.[e] When they moved, they went toward any one of four directions; they did not turn aside when they went. As for their rims, they were so high they were awesome; and their rims were full of eyes, all around the four of them. When the living creatures went, the wheels went beside them; and when the living creatures were lifted up from the earth, the wheels were lifted up." – Ezekiel 1:15-19 NKJV

Ezekiel's description of this wheel within a wheel sparked Blumrich to dive deeper. This wheel can move straight while also moving sideways

[e] Other Bible versions use the phrase "a wheel within a wheel."

without turning the craft. Once Blumrich discovered how this wheel operated, he used this knowledge to patent the Omnidirectional wheel, as noted in Figure 29.

Ezekiel notes that the wheel's rims were full of eyes. The eyes could have been small, short protrusions with inverted semicircles to aid the wheel's traction or tread. Today, this is seen at construction sites with the sheepsfoot roller. This machine's numerous protrusions help with compacting soil or wet clay.[6]

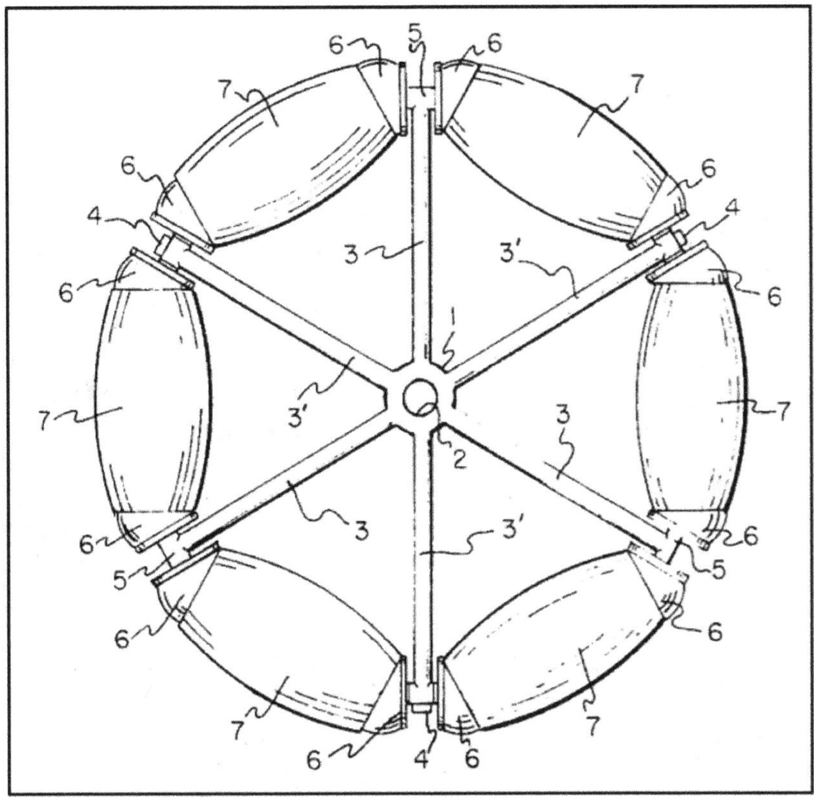

Figure 29: Blumrich's Omnidirectional Wheel Patent # #US3789947A

Credit: https://worldwide.espacenet.com/patent/search/family/022923089/publication/US3789947A?q=p

Ezekiel says the wheel's color is beryl, a type of mineral whose color ranges from emerald to aquamarine. Other Bible versions use 'chrysolite,' 'topaz,' or 'the sea.' Depending on the translation, they are described as 'gleaming' or 'sparkling.' This same type of coating is used today as a synthetic corrosion protection over machine steel and aluminum.[12] These wheels are also deployable independently from the landing feet.

The Craft's Lower Shape

Roger A. Anderson of NASA's Langley Research Center developed the ideal configuration of a conical shape for spacecraft. This shape is ideal because of the high aerodynamic drag and stability with low structural weight.[13] The lower part of the craft is similar to a type of spinning top, while the top of the craft is a curved dome. Encased in the convex underside are the reactor, radiator, plug nozzle, propellant, and radiation shield.[14] Ezekiel explains the rest of the description of the craft's body:

"The likeness of the firmament above the heads of the living creatures was like the color of an awesome crystal, stretched out over their heads."
– Ezekiel 1:22 NKJV

This verse gives a description of the craft with the four-column helicopters. Ezekiel uses "awesome crystal" to describe the color of the lower part of the craft. Other versions use 'terrible crystal,' 'terrible ice,' and 'rock crystal.' Many versions note the shiny or shimmering quality of the surface, which is similar to that of an alloy's smooth and shiny surface.[15]

The Command Capsule and Occupant

"And above the firmament over their heads was the likeness of a throne, in the appearance like a sapphire stone; on the likeness of the throne was a likeness with the appearance of a man high above it. Also from the appearance of His waist and upward I saw, as it were, the color of amber [*chasmal*] with the appearance of fire all around within it; and from the appearance of His waist and downward I saw, as it were, the appearance of fire with brightness all around. Like the appearance of a rainbow in a cloud on a rainy day, so it was the appearance of the brightness all around it. This was the appearance of the likeness of the glory of the Lord."

– Ezekiel 1:26-28 NKJV

Ezekiel noted his background and general knowledge at the beginning of chapter one. He would have been familiar with a throne. He could, however, be referencing an upholstered chair commonly seen today on airplanes and movie theatres. This chair was, as Ezekiel notes, the color of sapphire stone or royal blue. On the throne was the commander with "a likeness with the appearance of a man":

- Likeness = [דְּמוּת] *demuth*, figure, resembling
- Appearance, vision = [כְּמַרְאֶה] *maré* ,objectively observed
- Man = [אָדָם] *adam*, human

The figure that looked objectively like a human was described as the top portion of their body, which was a chasmal color with the appearance of fire around them. The lower portion of their body also appeared as fire and brightness. Ezekiel is not saying that the commander is on fire, but instead is possibly referring to Enoch's description of the Elohim. Chapter fourteen described the Watchers and Elohim in *The Book of the Secrets of Enoch*. He described their appearance as faces that "shone like the sun, and their eyes were like burning lamps"[16] and noted that their garments appeared as "a burning flame."[17]

Ezekiel calls the commander's appearance as a "rainbow in a cloud on a rainy day, with the appearance of brightness all around." Other versions note a glow, brightness, or light surrounding the commander. This could be another description of the Shining Ones noted by O'Brien, as explained in chapter five.

Ezekiel's Abductions

Ezekiel Chapter Eight begins with Ezekiel seated with the elders of Judah. The "hand of the Lord" comes upon him and he sees a Shining One. This "hand of the Lord" is mistranslated and portrays something mystical and weird. The Hebrew explanation shows something objective is happening around Ezekiel:

- Hand = [יָד] *yad*, comes from the root YD, 'hand moves.' 'work,' or 'gesture with hand.'[18]
- Of the Lord = [אֲדֹנָי] *adonay[i]*, from the root DN, 'door of life,' 'rule,' or 'the goal of the ruler is to guide their people.'[19]

"He stretched out the form of a hand [*yad*] and took me by a lock of my hair; and the Spirit [*ruach*] lifted me up between earth and heaven, and brought me in visions [*maré*] of God to Jerusalem, to the door of the north gate of the inner court, where the seat of the image of jealousy was, which provokes to jealousy." – Ezekiel 8:3 NKJV

- He stretched out = [שָׁלַח] send,' 'sent a messenger.'
- By a lock = [צִיצִת] *tsitsith* unknown origin and consists of the root *tsiyts*, 'wing-like projection,' or 'shining thing' (*kanaf*).[20]
- Of my hair = [רֹאשׁ] *rosh*, from the root RSh, 'representative' or 'ruler.'
- The seat =[מוֹשָׁב] *moshasb*, from the root ShB, 'returning' or 'turning back to a place for rest.'[21]
- Of the image of =[סֶמֶל] *semel*, uncertain origin, unused root meaning 'resemble.'
- Jealously = [קִנְאָה] *qinah*, from the root QN, 'gathering materials to provide protection over those whose well-being they are charged with.'[22]
- Which provokes to jealously = [קָנָה] *qanah,* from the root QN.

Retranslated: A messenger of Yahweh gestured and took me to the [*kanaf*] of the Elohim; and the [*ruach*] lifted me up between the sky and land,[f] and brought me [*maré*] to the Elohim, to the door of the north gate of the inner court, where he rested and gathered materials to provide protection.

"Then the Spirit [ruach] lifted me up and brought me to the East Gate of the Lord's house, which faces eastward; and there at the door of the gate were twenty-five men, among whom I saw Jaazaniah the son of Azzur, and Pelatiah the son of Benaiah, princes of the people." – Ezekiel 11:1 NKJV

"Then the Spirit took me up and brought me in a vision by the Spirit of God into Chaldea, to those in captivity. And the vision that I had seen went up from me." – Ezekiel 11:24 NKJV

- And went up = [וַיַּעַל] *alah*, from the root GhL, 'experience the staff,' 'yolk,' 'when those in exile are placed in the yolk.'
- From me = [עַל] *al*, same meaning as *alah*.[23]

While these may be very loose translations, it is easy to see that Ezekiel's experience had nothing to do with mystical visions the Church normally attributes to Ezekiel and that he is, in fact, just traveling back and forth from his home to speak with the Elohim. Ezekiel then travels with the messenger to Yahweh's house, where he sees 25 people, including the princes Jaazaniah and Pelatiah. He then returned to the *ruach* and was taken to Chaldea to see those in captivity. He objectively saw the yolk of those in captivity.

---

[f] Chapter five notes translation of 'Heavens' [הַשָּׁמַיִם] *haš-šā-ma-yim,* meaning 'sky' or 'skies,' and Earth [הָאָרֶץ׃] *hā-'ā-res,* meaning 'land.'

Other Visions of Sky Chariots

UFOs have been discussed throughout history. As the stories of Elijah and Ezekiel show, they were commonly referred to as 'sky chariots' or 'chariots of fire.' Titus Livius, known as Livy, was a Roman historian who wrote the *History of Rome*. In 218 BCE, he noted that ships were brightly visible in the sky. In 217 BCE, round shields appeared in the sky over Italy, and in 173 BCE, a great fleet was seen in the sky.[24]

The Apocrypha text of 2 Maccabees also describes when UFOs were seen in the sky around 100 BCE:

"And it so befell that thought all the city, for the space of almost forty days, there appeared in the midst of the sky horsemen in swift motion, wearing robes inwrought with gold and carrying spears, equipped in troops for battle; and drawing of swords; and on the other side squadrons of horse in array; and encounters and pursuits of both armies; and shaking of shields, and multitudes of lances, and casting of darts, and flashing of golden trappings, and girding on all sorts of armor." – 2 Maccabees 5:2-3

In his book *Wars of the Jews*, Josephus describes seeing UFOs during Rome's attack on Jerusalem in 65 CE. He was so baffled by this phenomenon that he had to specify and state that he would not have believed it if not for the people who told him:

"Thus there was a star resembling a sword, which stood over the city, and a comet, that continued a whole year...a certain prodigious and incredible phenomenon appeared; **I supposed the account of it would seem to be a fable, were it not related by those that saw it**, and were not the events that followed it of so considerable a nature as to deserve such signals; for before sunsetting, *chariots and troops appeared in their armor were seen running about among the clouds.*"[25] (emphasis added)

. . .

Lastly, Plutarch of Chaeronea, who lived 45-120 BCE, was a Greek writer who wrote about the life of a Roman General named Lucullus in his book *Parallel Lives Vol. II*. Lucullus and his army were on the verge of battle against the Mithridates at Otryae when the sky opened up and a mysterious object appeared and came between both armies. Lucullus notes that it had the shape of a cask and had a molten silver color.[26]

While these are only a few examples of the 'chariots of fire' seen throughout history, Ezekiel's description of his encounter with the chariot of fire specifically explains UFO contact instead of mystical flying chariots that are also on fire.

Discovering Yahweh: The Bible and Stories of the Anunnaki

Throughout Part II, it may have become apparent that there were alterations in the Bible's cohesive narrative, and the truth of extraterrestrial contact was concealed. Previous chapters correlated biblical characters and stories to the stories of Egypt and the important role Egypt held in the ancient world. This chapter will compare the Bible creation story with the narratives of the Anunnaki, highlighting the potential skewing of the Bible stories and providing a broader understanding of other ancient tales from Mesopotamia to the Americas.

Sitchin accurately explains the Genesis creation stories. The level of detail in as few words as possible shows their mastery of precision.[1] The story of the Anunnaki parallels almost perfectly with *The Terra Papers*. Whether or not the two stories are the same, which is highly probable, they complement one another exceedingly well. The story's beginning was told in chapter four and will not be repeated, but the correlations between the stories are in the footnotes to avoid confusion.

The Anunnaki story on Earth begins when their planet, Nibiru, was in decline. They discovered that gold was needed to save their planet and species. About 450,000 years ago, Alalu sat on the throne of Nibiru with his cupbearer Anu.[a] AN.U's (Heaven, Great Father of the Gods)[b] father AN.SHAR.GAL (Great Prince of Heaven) was heir apparent with his half-sister wife KI.SHAR.GAL (Great Princess of Firm Ground). In the epic *Kingship in Heaven*,[2] Alalu was king for seven years before Anu battled him for the throne. After Anu won, Alalu fled to Earth. Anu sent E.A (Whose House is Water) to Earth to begin mining for gold.[c] Ea arrived with his group of 50 Anunnaki (Those Who From Heaven to Earth Come)[d] in their DIN.GIR (Righteous Ones of Rocketships) on Earth 432,000 years before the Deluge. [3] Table

GENESIS CHAPTER 1

Day One and Two

"In the beginning, God[e] created the heavens and the earth. The earth was without form, and void; and darkness was on the face of the deep. And the Spirit of God was hovering over the face of the waters. Then God said, 'Let there be light,' and there was light. God called the light

---

[a] *Terra Papers* notes that King AN-AN was the Elder king of ASA-RRR. Princes and other possible heirs were expected to plot the demise and usurp the throne. The Royal Cupbearer of An-An was AL-SHAR, who successfully stole the throne. AN-U, son of AN-AN, was made the royal cupbearer to the new king, AL-AL-IM (the name he was given after becoming king). AN-U served as his royal cupbearer, and when he had the chance, he took back the throne and allowed AL-AL-IM to control the solar system (Morningstar. *Terra Papers*, p. 8-10, 12). The ASA-RRR were already aware of the solar system and helped develop it. They originally settled Tiamat but it exploded in the epic space battle of An-u against Zu-zu. The dates given are based on Sitchin's timeline in *The Wars of Gods and Men*, p. 345-350, and *The 12th Planet*, p. 410-11.

[b] All Anunnaki names are spelled specifically to convey the meaning of the character.

[c] Sitchin notes that he was a brilliant scientist and engineer with the epithet NU.DIM.MUD (*The Wars of Gods and Men*, p. 78.) *The Terra Papers* state Ea was also a master Genesis scientist who would conduct mining and metallurgy operations on Earth (*Terra Papers*, p. 14.).

[d] Morningstar notes that the Anunnaki was the name given to the volunteers from the home planet ASA-RRR. These volunteers were each given the rank of Lord along with property on Eridu (the first name of Earth) and a monetary stake in the new planet (Ibid, p. 27.).

[e] The footnote of the word God, translate as Elohim, the plural form of Eloah.

Day, and the darkness He called Night. So the evening and the morning were the first day.

"Then God said, 'Let there be a firmament in the midst of the waters, and let it divide the waters from the waters.' Thus God made the firmament, and divided the waters which were under the firmament from the waters which were above the firmament; and it was so. And God called the firmament Heaven. So the evening and the morning were the second day." – Genesis 1:1-7 NKJV

God had already made the heavens and earth. He then divided the fresh water from the salt water. Once he was finished with the water, he created darkness and light, thus creating the first true day. The translation in chapter five provides a clearer picture of what occurred: initially, the Elohim shaped the sky and the land. However, the land eventually descended into primeval chaos. Darkness covered the surface of the water, and the UFO of the Elohim hovered, suspended above the water's edge.[f]

Ea found a place to land and immediately established an outpost station named E.RI.DU (House in Faraway Built) near the region of the Persian Gulf and marshlands. Gold production began to slow down and more Anunnaki were sent in groups of 50 to assist. Approximately one Nibiru orbit cycle (1 Nibiru year = 3600 Earth years) after Ea landed, NIN.MAH (Great Lady) and MAR.DUK (Son of the Pure Mound) arrived. Ice sheets began to recede, and the earth became more hospitable. Ninmah set up a medical center, and the Earth station was able to move inland, establishing Larsa (also called Larak). Easily obtainable gold was becoming more difficult to find.

Ea discovered that gold was abundant in a specific region, but they needed to mine it, thus the AB.ZU (The Primeval Source) was estab-

---

[f] Morningstar writes of Ea's spaceship as it moved slowly over the land and examined the surface of the destroyed world (*The Terra Papers*, p. 27.).

lished in Africa. After gold production exploded into colonies of gold mining, more resources, facilities, transport, communications, and jobs were needed to continue—just like business expansion runs today.[3]

Day Three, Four, and Five

"Then God said, 'Let the waters under the heavens be gathered together into one place, and let the dry land appear,' and it was so. And God called the dry land Earth, and the gathering together of the waters He called Seas. And God saw that it was good. Then God said, 'Let the earth bring forth grass, the herb that yields seed, and the fruit tree that yields fruit according to its kind, whose seed is in itself, on the earth; and it was so. And the earth brought forth grass, the herb that yields seed according to its kind, and the tree that yields fruit, whose seed is in itself according to its kind. And God saw that it was good. So the evening and the morning were the third day.

"Then God said, 'Let there be lights in the firmament of the heavens to divide the day from the night; and let them be signs for seasons, and for days, and years; and let them be for lights in the firmament of the heavens to give light on the earth;' and it was so. Then God made two great lights: The greater light to rule the day, and the lesser light to rule the night. He made the stars also. God set them in the firmament of the heavens to give light on the earth, and to rule over the day and over the night, and to divide the light from the darkness. And God saw that it was good. So the evening and the morning were the fourth day.

"Then God said, 'Let the waters abound with an abundance of living creatures, and let birds fly above the earth across the face of the firmament of the heavens.' So God created great sea creatures and every living thing that moves, with which the waters abounded according to their kind, and every winged bird according to its kind. And God saw that it was good. And God blessed them, saying, 'Be fruitful and multiply, and fill the waters in the seas, and let birds multiply on the earth.' So the evening and the morning were the fifth day." – Genesis 1:8-23 NKJV

On the third day, God separated the various forms of water and brought about land. He created vegetation on the land and caused the vegetation to yield more vegetation. The fourth day seems like a repeat of the second day when light, aka the sun, was already made. It is not like this god hovered over water and turned on a light switch; the sun was present before the planet formed. The moon arrived and gave light to the night. The stars were also created, even though there are stars older than the age of this solar system. On the fifth day, marine life and birds were created.

After 28,800 years of success, Anu, father of Ea and ruler of Nibiru, decided more Anunnaki were needed. While Ea was in charge of production, Anu would place his son EN.LIL (Lord of the Command) in charge of logistics. This act by Anu is one of many decisions that led to rivalry between the two brothers. The *Atra-Hasis* outlines Anu's attempt to calm this rivalry by casting lots to decide. Anu was to govern heaven, Enlil governed the airspace, and Ea/Enki was given the land and seas.[g,3]

The arrival of Anu, Enlil, and the rest of the Anunnaki was around the time of the interglacial period, approximately 400,000 years before the Deluge. Enlil established Nippur as the Mission Control Center. The Spaceport was established at Sippar, and Shuruppak became the Medical Center. Ea/EN.KI (Lord of the Earth) established a metallurgic center at Bad-tibira to refine ores. The refined gold was transported to the orbiters manned by the IG.IGI (Those Who Observe and See) and sent to Nibiru.[h 4]

---

g When Ea was given dominion over the land, he was granted the title EN.KI (Lord of the Earth).

h *The Terra Papers* notes that the pyramid was built to create stability for the energy grid. It was called the "energy house" and was the location of the "re-animation chamber" where Ninmah was. Eridu's Agricultural-Biological Center was the main hub, and it was

Anu left after the lots were cast and returned to Nibiru. Everything was going smoothly. Back in Nibiru, Anu kept his enemies close, and Kumarbi, the son of Alalu, was made his Royal Cupbearer.[i] After eight years (28,800 earth years), the Anunnaki told Anu that the Igigi were getting upset because the work was too hard. A year later (3600 Earth years), Anu returned to Earth with Kumarbi for this second visit. During this visit, the decision was made to create primeval man to replace the Igigi's workload. While returning to the ship, Kumarbi revolted, and Anu fled to the sky but was caught by Kumarbi, and Kumarbi bit Anu's genitals.[j] Kumarbi went to Ea for guidance and was directed to speak with the king in heaven (Lamma). The celestial gods revolted against the king in heaven. Kumarbi stayed and, with the help of the Igigi, waged a battle against Tessub, the storm god. Kumarbi impregnates the mountain, and a giant named Ullikummi is born. Before Kumarbi dies, he tells his son to hide until he is grown and then to avenge his death. This story becomes what is today known as the War of the Olympus Gods (the celestial gods versus the gods of the netherworld versus the titans). Tessub is victorious, and Anu returns to the throne. This myth is also told in the *Myth of Zu*. Kumarbi is Zu who has stolen the Tablets of Destiny, and Ninurta must kill him and take them back to claim his Enlilship.[k,5]

. . .

---

starting to look more like a city than a short-term settlement. Ea worked to engineer hybrid animals and plant seedlings in his laboratory that could work to transform the atmosphere. He transported them across the globe, and it caused the planet to begin warming. To increase production, EA created the H-N lizard hybrid to assist in underground mining operations (known as HEN-T Hybrids) (Morningstar 28-29).

i King AL-AL-IM's son, AL-AL-GAR, was born to one of the IKIKI pilots (the orbiting Watchers of Anu) and achieved the highest status among pilots. He trained along the elite warriors of the BEH forces in secret. He achieved the title of Master Supreme and was renamed ZU-ZU. King AN-U placed him as his Royal Cupbearer to avoid history repeating itself. Threats from the IKIKI pilots reached AN-U, and he took ZU-ZU with him to quell the uprisings in his planet ship (AR). When AN-U returned, ZU-ZU attacked with the IKIKI, and AN-U was sent back to ASA-RRR; ZU-ZU became ruler of the region (Morningstar p. 14-17).

j Reference is made to the similarities between these stories. Yet, *Terra Papers* notes that Anu's genitals (DAK was a highly skilled warrior) and reference to Anu's testicles (Anu's ball of power [AR]) (Morningstar 17).

k The War against ZU-ZU and AN-U ends with one death planet ship versus the other

Day Six

"Then God said, 'Let the earth bring forth the living creature according to its kind: cattle and creeping thing and beast of the earth, each according to its kind'; and it was so. And God made the beast of the earth according to its kind, cattle according to its kind, and everything that creeps on the earth according to its kind. And God saw that it was good.

"Then God said, 'Let Us make man in Our image, according to Our likeness; let them have dominion over the fish of the sea, over the birds of the air, and over the cattle, over all the earth and over every creeping thing that creeps on the earth.' So god created man in His own image; in the image of God He created him; male and female, He created them. Then God blessed them, and God said to them, 'Be fruitful and multiply; fill the earth and subdue it; have dominion over the fish of the sea, over the birds of the air, and over every living thing that moves on the earth.

"And God said, 'See, I have given you every herb that yields seed which is on the face of all the earth, and every tree whose fruit yields seeds; to you it shall be for food. Also, to every beast of the earth, to every bird of the air, and to everything that creeps on the earth, in which there is life, I have given every green herb for food,' and it was so. Then God saw everything that He had made, and indeed it was very good. So the evening and the morning were the sixth day." – Genesis 1:24-31 NKJV

On day five, the earth's animals were made, particularly the cattle and the creeping things, even platypuses that creep on the earth (maybe

---

death planet ship, the destruction of TIAMAT, and the ultimate capture and murder of ZU-ZU at the demand of the ARI-AN queens (Morningstar 18-25).

those were made on the fourth day). He then created man and woman in his own image—even the Bible says humans look just like God—and says everything was created for them for food.

The Anunnaki storyline continues around 270,000[l] years before the Deluge, with the formation of primeval man told in *Enki and Ninmah*. In the first attempt, Ninmah got drunk while making the primeval man, six deformed beings were made, and the blame was placed on Enki.[m] The creation of man in the *Atra-Hasis* addresses a possible second attempt. Belet-ili, the womb goddess and midwife of the gods,[n] worked with Enki to create man.[o] Clay was pinched off, and a sacrifice was made.[p] Ninmah mixed the clay (the flesh of man) was mixed with blood (the blood of the gods, "divine blood"). Ninki,[q] the wife of Enki, carried the child in her womb and then gave birth. The drumbeat was heard, and the ghost (spirit) came into existence. The first human man was the perfect imprint of the gods. He was named Adapa, and he was known as

---

[l] Genetic studies note that the genetic markers for the DNA of *Homo sapiens* are only through the female's mitochondrial DNA. These studies indicate that there was a single Eve approximately 200,000 years ago. The origin of the Y chromosome points to a single Adam that lived 340,000 years ago. Sitchin's books were published in the late 1990s, and genetic studies done since show earlier times than Sitchin initially placed in his works. Yet the single Y chromosome ancestor was said to be an ancient African human. Sitchin suggests that this ancient African human was Neanderthal (Sitchin. *Divine Encounters*, p. 45; Mendez, Fernando L., et al. "An African American Paternal Lineage Adds an Extremely Ancient Root to the Human Y Chromosome Phylogenetic Tree,"*American Society of Human Genetics*, February 28, 2013. https://doi.org/10.1016/j.ajhg.2013.02.002).

[m] *The Terra Papers* also tell of odd creatures that were created. NIN-HUR-SAG and Ea instead used the genetic material of the ASA-RRR species and mixed it with ERIDU beasts. They already had seen success with the H-N hybrid lizard they created and worked to create more worker creatures. A centaur and a minotaur were created and were successful. They kept working and created the ape beast named APA. This creature had incredible strength but little intelligence and was perfect for the hard work needed (Morningstar 31-32).

[n] Another name for Ninmah.

[o] Out of anger for not being chosen as AN-U's successor, Ea decided to adjust the APA creature out of revenge towards AN-U and EN-LIL. Ea used his own DNA and worked through prototypes to create ADAPA (Morningstar 33).

[p] *Atra-Hasis* notes a god was slaughtered, Sitchin notes that blood was poured into the tub the clay was mixed in and no god was slaughtered.

[q] NIN.KI (Lady of the Earth), also named Damkina before she married Enki.

the offspring of Enki because his wife carried him. He became the model from which all other humans would be copied (cloned).[6]

Genesis Chapter Two

"Thus the heavens and the earth, and all the host of them, were finished. And on the seventh day, God ended His work which He had done, and He rested on the seventh day from all His work which He had done. Then God blessed the seventh day and sanctified it, because in it He rested from all His work which God had created and made. This is the history of the heavens and the earth when they were created, in the day that the Lord God[r] made the earth and heavens." – Genesis 2:1-4 NKJV

A mist went up from the earth to water the whole planet before the plants and animals were created. There was no rain and no one to work the fields, so God created man from the Earth, and he became a living being. – Genesis 2:5-7 summarized

These verses are a reprieve from the first chapter in Genesis, but it is interesting to note the specifics of "no one to work the land." It is as if the entire purpose of creating everything was to make man work unless everything was created, and then workers were needed because god was not able to create something that did not need maintenance work. As the Anunnaki story shows, this was the purpose of creating man.

The Garden of Eden

"The Lord planted a garden eastward in Eden, and there He put the man whom He had formed. And out of the ground the Lord God made

---

[r] This is the first time there is a separation between God (Elohim) and Lord God (Yahweh Elohim).

every tree grow that is pleasant to the sight and good for food. The tree of life was also in the midst of the garden, and the tree of knowledge of good and evil.

"Now a river went out of Eden to water the garden...the first is Pishon, where there is gold, and the gold of that land is good. Bdellium and the onyx stone are there... Gion; it is the one which goes around the whole land of Cush... the third river Tigris that goes east of Assyria. The fourth river is the Euphrates.

"The Lord God took the man and put him in the garden of Eden to tend and keep it. And the Lord God commanded the man, saying, 'Of every tree of the garden you may freely eat; but of the tree of the knowledge of good and evil you shall not eat, for in the day that you eat of it you will surely die." – Genesis 2: 8-17 NKJV

The verses thus far reveal a lot of information in a very precise and unassuming way. Previous Anunnaki stories have already shown that Adam was specifically created to work; whether primitive or not, he was made to ease the workload of the Igigi. The Hebrew text uses *adam* (אָדָם) for man.[s] The word's origin is unknown and translates as 'anyone,' 'common sort,' 'infantry,' 'hypocrite,' or 'low degree.' The Hebrew use of the word "man" is an inaccurate transliteration of Sumerian and Akkadian texts that note *lulu* (primitive), *awilum* (laborer), and *lulu amelu* (primitive worker). Another inaccurate transliteration is 'worship.'[t] The Sumerian term used was *avod* (work).[7]

---

[s] Strong's Concordance No. 120.

[t] *The Terra Papers* note that EN-LIL did not trust the ADAPA beast despite a strong demand for him. He was forced by AN-U to use them and, therefore, decided to place the beast in the most difficult working environment possible. The beasts were clones and, thus, expendable. If the ADAPA beast became sick or died, it was meaningless to EN-LIL, for they were just beasts. Prince Ea realized that after all his hard work to create ADAPA,

The *Atra-Hasis* echoes this after the creation of the first man. Ninmah tells the Anunnaki and Igigi that she relieved the gods of their hard work and imposed it upon man. She gave the gods freedom by placing the workload on man.[8] In ancient times, the word 'Lord' was synonymous with 'Master,' 'Sovereign,' and 'Ruler.' Man did not worship the Lord; he was a primitive laborer who worked for the Master. And what was this *lulu amelu*'s first task? To tend to the garden and collect gold, of course. He was shown all the wondrous trees and told which ones he could eat from and which ones were forbidden, with the consequence of disobedience being death.

The Creation of Woman

"And the Lord God said, 'It is not good that man should be alone; I will make him a helper comparable to him.' Out of the ground the Lord God formed every beast of the field and every bird of the air, and brought them to Adam to see what he would call them...But for Adam there was not found a helper comparable to him.

"And the Lord God caused a deep sleep to fall on Adam, and he slept; and He took one of his ribs, and closed up the flesh in its place. Then the rib which the Lord God had taken from the man He made into a woman, and He brought her to the man. And Adam said: 'This is now bone of my bones and flesh of my flesh; She shall be called woman, because she was taken out of man.' Therefore a man shall leave his father and mother and be joined to his wife, and they shall become one flesh. And they were both naked, the man and his wife, and they were not ashamed." – Genesis 2:18-25 NJKV

---

they had become inconsequential slaves to EN-LIL and not the laborer helpers that Ea had intended (Morningstar 33).

How does the first man, supposed to know nothing, understand what a husband and wife are? What about a mother and father and ceremonial rights after marriage? How did Adam and Eve know that they were naked? If they walked with the Lord God around Eden, was God naked? Or was Lord God clothed, and they understood that there was something different but were unashamed?

The Anunnaki's story reveals the creation of a second pair of males and females approximately 250,000 years before the Deluge. Fourteen pieces of clay (flesh) were separated, seven on the right and seven on the left, and a mudbrick was between them. Ninmah then used a reed to cut the umbilical cord and, with artificial insemination, placed the clay,[u] divine blood mixture, and likeness (soul) into the assembly of womb goddesses. After ten months of waiting, the staff slipped in, and the womb was opened. She had created male and female. After the success, Ninmah told the Anunnaki what she had made and said a wife and husband would choose each other. [9] After the creation of male and female, Ninmah was given the name NIN.TI with TI meaning 'life' and 'rib.'[10] Thus, the woman was not made from Adam's rib, but rather, the one who created them both had the epithet "Lady of the Rib."

GENESIS CHAPTER THREE

The Fall of Man

"Now the serpent was more cunning than any beast of the field which the Lord God had made. And he said to the woman, 'Has God indeed said, 'You shall not eat of every tree of the garden'?' And the woman said to the serpent, 'We may eat the fruit of the trees of the garden; but of the fruit of the tree which is in the midst of the garden, God has said, 'You shall not eat it, nor shall you touch it, lest you die.' Then the serpent said to the woman, 'You will not surely die. For God knows that in the

---

[u] The Akkadian term for clay is *tit,* which translates to TI.IT in Sumerian, meaning 'that which is with life.' The Hebrew *tit* root is synonymous with that of egg (Sitchin. *The 12th Planet*, p. 357).

day you eat of it your eyes will be opened, and you will be like God, knowing good and evil.' So when the woman saw that the tree was good for food, that it was pleasant to the eyes, and a tree desirable to make one wise, she took of its fruit and ate. She also gave to her husband with her, and he ate. Then the eyes of both of them were opened, and they knew they were naked; and they sewed fig leaves together and made themselves coverings." – Genesis 3:1-7 NKJV

They heard the sound of the Lord God walking in the garden and hid. The Lord God asked where they were and Adam said that they hid because they were afraid because they were naked. Lord God asked who told them they were naked and if they had eaten from the tree. Adam said that Eve was the one who ate from the tree and then gave it to him. The Lord God got mad. Eve said that the serpent deceived her. The Lord God cursed the serpent with hostility between the serpent and the woman, the seed of the serpent and the seed of the woman. The Lord God cursed Eve and gave her pain during childbirth, and the man ruled over her. The Lord God cursed Adam to labor all throughout his life and cursed him to die. The Lord God made tunics of skin to wear as clothes. – Genesis 3:8-21 (summarized)

"Then the Lord God said, 'Behold, the man has become like one of *us*, to know good and evil. And now lest he put out his hand and take also of the tree of life, and eat, and live forever'—therefore the Lord God sent him out of the garden of Eden to till the ground from which he was taken." – Genesis 3:22-23 NKJV (emphasis added)

The story of the fall of man is the thing that has separated mankind from God. It is the tool that the church has used since the church was started to explain why humans *need* God. It is also taught that Eve was the one who had to talk Adam into eating the fruit as if he was somewhere else when Eve ate the apple, yet the verse notes that Adam was right next to Eve when the serpent was present. But is this story true?

Not only does Sitchin tell a different story but the *Apocrypha*, *Nag Hammadi*, and books from the *Dead Sea Scrolls* all tell different stories. Sitchin begins by noting that when Adam and Eve were created, they could not procreate. He even notes that Adam and Eve were not modern-day humans but the progenitors of *homo sapiens*.[11] The *Book of Jubilees* tells the story of when Adam and Eve were in the Garden of Eden:

"[God] showed [Eve] to [Adam]: and for this reason, the commandment was given to keep in their defilement, for a male seven days, and for a female twice seven days. And after Adam had completed forty days in the land where he had been created, we brought him into the Garden of Eden to till and keep it, but his wife they brought in on the eightieth day, and after this, she entered into the Garden of Eden...and in the first week of the first jubilee, Adam and his wife were in the Garden of Eden for seven years tilling and keeping it, and we gave him work, and we instructed him to do everything that is suitable for tillage. And he tilled (the garden), and was naked and knew it not, and was not ashamed, and he protected the garden from the birds and beasts and cattle, and gathered its fruit, and ate and put aside the residue for himself and for his wife." – Jubilees 3:8-9, 15-16

They were kept in confinement areas, and Adam was in charge of keeping up the garden. They were naked slaves given scraps of food, or the residue, to eat after the gods had their share. The story states that Adam was to work the land for seven years. The serpent came, Eve ate the fruit, then gave it to Adam, and all flesh was sent out of the garden. After they left the garden,

"They had no son until the first jubilee, and after this he knew her. Now he tilled the land as he had *been instructed* in the Garden of Eden." – Jubilees 3:34-35 (emphasis added)

. . .

They were not allowed to procreate until after they left the Garden of Eden. While most people assume their first son was Cain, this is not the case—but let's not jump too far ahead yet. Adam was instructed to till and care for the Garden. He was allowed to eat the scraps from all the trees but only to look at the Tree of Knowledge of Good and Evil, as the Bible puts it, and not even allowed to touch it. He walked around naked and was only given clothes after they left. Jubilees notes that only Adam and Eve were given clothes, and none of the "other flesh" expelled from the garden received clothes. This leads to more questions. What is the other flesh? Why was God so strict about a tree, and if it was so bad for man to touch, why even put it in a garden and let them access it? Thankfully, the "Revelation of Adam" text gives the answers.

Adam tells Seth that when he and Eve were created out of the earth, they were in the eternal realm, in glory. Eve taught him the knowledge of god: that they resembled the angels and were greater than the god who created them. They were even greater than the powers of a god they did not know, and this unknown god-ruler of the realms divided them in anger. Their glory was gone, along with the knowledge, and a new eon began. Adam remarks that they came from another generation that knew the great eternal beings. Now that they were human, their knowledge was gone, and they only learned about mortal things, like humans.[v] They began recognizing the god who created them, and his powers became known. They served him in fear and servitude while their minds became dim.[12]

This perspective shows that there was no fruit, tree, or garden; there may have been a garden. Adam speaks of two different gods, the one who created them (Ea) and another one he did not know (Enlil). The Biblical story places Eve as the one who ate from the tree and then shared it with Adam. The "Revelation of Adam" says that Eve shared the knowledge with Adam that they were created as greater than the angels and even the gods themselves. As for the tree of life mentioned in Genesis, the three words used to describe the genetic engineering and

---

[v] The translation by Hedrick notes that they were only taught dead things (Hedrick. *Apocalypse of Adam*, p. 231).

manipulation of DNA that Ea did to create humankind are amazing. His symbol was the double-entwined serpents; thus, the double helix tree of life motif was created.[W]

As for the "other flesh" mentioned in Jubilees, there were multiple beings created by Ninmah, and Enki had created more beings in the AB.ZU who stood around him.[13] The other god who Adam and Eve did not know is also discovered in "The Creation of the Pickax." Enlil created the pickax and was very pleased with what he made. He showed it to the other gods, and while they oohed and awed over it, he handed it to the "black-headed people" as a gift. The text says that the pickax builds and helps things prosper, but for those who rebel or are not submissive to the king, the pickax makes them submissive. Enlil decreed the fate of the pickax and it was exalted.[14]

When the black-headed people were created by Ea, Enlil made the pickax and then handed it to the black-headed people. So Enlil was one of the other beings mentioned in the Bible who walked in the garden. Ea was the serpent who gave the knowledge to Adam and Eve, and Enlil punished them with the ax and cursed them to die because they were not submissive after they gained the knowledge.[X]

[W] This motif was also given to Ningishzida/Thoth (Sitchin. *Divine Encounters*, p. 14.).

[X] When Prince Ea discovered what EN-LIL was doing to his ADAPA beasts, along with how his brother had completely taken over everything he had worked so hard for, he decided to use his creation against EN-LIL. Ea knew the beasts were fed in the Garden. He knew there were strict breeding schedules (yes, like cattle), and only certain beasts were allowed to mate under specific conditions, and they were absolutely not allowed to mate without the approval of EN-LIL. Ea found a few that were alone and told them about spontaneous pleasure outside of the strict regulations. The beasts could now be like the masters, enjoy a moment of pleasure and intimacy without scheduling or approval, and know true happiness. He gave them "knowing." Ea's symbol was the double-entwined serpents. When EN-LIL found out, he was furious and kicked out all the offensive beasts, then established Commands of the Lord of the Word (similar to those of the Ten Commandments): 1) Total obedience, 2) No other remembrances of Ea and only total obedience, 3) No evil sounds or utterances about EN-LIL, 4) Attend obedience lessons every seventh period, 5) No mating outside of the approved schedule. The ADAPAS became the ADAMUS (MUS – monsters), and the females were to suffer childbirth because their access to birthing chambers was revoked. The ADAPAS were the faithful, and the ADAMUS were the unfaithful group who were sent to die in the wilderness. Ea still wanted to do more for the ADAMUS. He wanted to give them intelligence and the

GENESIS CHAPTER FOUR

Cain and Abel

"Now Adam knew his wife, and she conceived and bore Cain, and said, 'I have acquired a man from the Lord.'"[y] – Genesis 4:1 NKJV

After the initial reading, this verse sounds like Eve conceived Cain with Adam, and because of the grace of the Lord, she bore Cain. Yet this is not what this verse is saying. Look at it again. "I have acquired a man *from* the Lord." The "Revelation of Adam" reveals when Adam was asleep, three people appeared to him that he could not recognize because they were not produced by the powers of the one who created him (Adam now saw Enlil as the one who created him and forgot about Ea).[z] These three people surpassed glory and told Adam and Eve about the eternal realm and how life came about—aka that Ea created them— and who taught them the wisdom they were made to forget—again, Ea. Both Adam and Eve remembered and called Ea the God who created them and received the soul into them. Then, Ea created a son from himself and Eve. Out of this creation, the desire for Eve came along with remembering that their knowledge was gone and they were bound to death.[12] This story shows that Cain was the son of Eve but not Adam

---

freedom and independence to act outside of the system imposed by ENLIL. Ea developed the speech of ADAMUS and worked with NINHURSAG to teach wilderness survival, clothes making, and the understanding of symbols. The ADAMUS that showed a higher ability to learn and teach knowledge to other ADAMUS became known as the EA-SU (They who knew the way of Ea) (Morningstar 33-36).

[y] This is the first time "Lord" is used by itself. It appears in texts as LORD [יְהוָה] *Yahweh*, the proper name of the God of Israel. Strong's Concordance no. 3068.

[z] *The Terra Papers* notes that it was Ea and NINHURSAG who went to the ADAMUS after their expulsion from Eden. Sitchin notes that the first Adapa being lived with him in Eridu; the three beings are possibly Ea, Ninhursag (Ninmah), and the first Adam, who still had immortality because it was never stated that this Adapa died. The "First Stele of Seth" notes that Seth refers to Pigeradamas, the heavenly Adam, which could correlate to the first Adam who lived with Ea (Meyer. *Nag Hammadi*, p. 523).

and was essentially created via modern-day techniques—i.e. invitro fertilization.

The First and Second Books of Adam and Eve give their experiences after leaving the garden. They walked into the wilderness from a luscious, forested area (Eden). God made a covenant with Adam, promising to send someone to save him and his descendants in 5,500 years. Though they mourned their situation, God reassured them to move on because help would eventually come. God then provided them with a Cave of Treasures, though no physical treasure was found in the cave. Adam did not want to enter the cave but knew that unless they went into it, they would be labelled "transgressors" again. Thus, a picture is painted of a psychopathic god who makes them feel bad due to their actions. However, God explained that it was Satan who made the forbidden tree appear, and then referred to Satan as having no good intentions and seeking to take His place in heaven. After that, Satan appeared to Adam many times over seven months and thirteen days.[15] If we compare this to *The Terra Papers*, which says that Ea went to the ADAMUS in the cave to help them survive, it could be that Satan is equivalent to Ea. Could these stories be similar?

Abel was born. Abel was a shepherd, and Cain was a tiller of the ground. Cain brought an offering of fruit to the Lord. Abel brought the firstborn of his flock. The Lord respected Abel and his offering but did not respect Cain and his offering. This made Cain angry. The Lord asked why Cain was angry and then told him that if he did well, he would be accepted. If he did not do well then sin lay at his door, but he should rule over his desire. Cain killed Abel and when the Lord asked where Abel was, Cain rudely told Him that he did not know and was not his keeper. The Lord found out that Abel was killed by Cain and cursed him from the earth to be a vagabond. Cain said that anyone who found him will kill him. The Lord put a mark on him and said that anyone who killed him will receive vengeance sevenfold.

. . .

Cain travelled to Nod, east of Eden. He lay with his wife, and she conceived Enoch. They built a city and named it Enoch. The patrilineal lineage was listed: Enoch, Irad, Mehujael, Methushael, and Lemech. Lemech had two wives, Adah and Zillah. To Adah was born Jabal (those who live in tents) and Jubal (those who play the harp and flute).[aa] To Zillah was born Tubal-Cain the craftsman of bronze and iron, and the sister Naamah." – Genesis 4:2-24 (summarized)

It is interesting to note that Cain was really disliked by the Lord but does not explain why. Cain worked just like Abel did. When Cain asked why the Lord did not accept his offering, he was told that if he did well, he would be accepted, and if he did not do well, then sin lay at his door, and that sin desires Cain. Even in *The First and Second Books of Adam and Eve*, Cain was given a name that meant 'hater' and consistently referred to as having a 'hard heart.'[16] This does not make any sense. Unless Enlil knew that Cain was born of Ea and Eve. Before Abel was killed, Satan approached Cain and appeared in the figure of a man. Satan told Cain that if he followed him, he would be better off than Adam.[16]

The Bible provides no information about Cain and his lineage besides this passage. Canaan is often called the land of Cain, but it was the land of Canaan because Noah's grandson Canaan moved into that region. The lineage and history of Cain are found across the globe in South America in the stories of the Nahuatl tribe. Sitchin writes of his lineage and notes that the city Cain built was called Enoch and is eerily similar to the Aztec capital Tenochtitlán (The City of Tenoch: T-Enoch). The Nahuatl tribe also tells of a wandering patriarch who built the city, his

[aa] It is of interest to note the similarities between the names of the two sons of Lemech and the very similar names attributed to the three people who killed Hiram Abiff (Jubela, Jubelo, and Jubelum) from chapter eighteen. Even more interesting are the various references throughout the Bible, all made towards seeking truth, even though it has been masked in secret teachings. It is as if those who usurped the real knowledge were also usurped by those who held the real knowledge and placed easter eggs of information within everything to point towards universal awakening and understanding that human beings are not the only ones in this universe. This topic will be discussed in Part III.

descendants who evolved into the tribal nations of South America, and the son who was a master craftsman of metals. Sitchin also remarks that Cain could have the fourth progenitor's skin color—red—as Shem, Ham, and Japhet are the origins of the Asian, African, and European races respectively. The mark made by God was the trait of Natives across America: the absence of facial hair. [17] The lineage of Cain thus beings the elusive Toltec Civilization.

The Birth of Seth

"And Adam knew his wife again, and she born a son and named him Seth, 'For God has appointed another seed for me instead of Abel, whom Cain killed.' And as for Seth, to him also a son was born; and he named him Enosh. Then men began to call on the name of the Lord." – Genesis 4:25-26 NKJV

In "The Three Steles of Seth," Seth states that his father was the heavenly Adam, referring to the first Adam created by Ea, and calls himself "Emmacha Seth." He reveals that he was born the same way as Cain but could procreate himself.[18] Genesis also reveals that it was only after the birth of Enosh, Seth's son, that *men* called on the name of the Lord, showing that Seth was possibly not a modern-day human. Sitchin speculates that Genesis explains the changes between Neanderthal (Adam), Cro Magnon (Seth), and *Homo sapiens sapiens* (Enosh) in a very precise manner. 'Enosh' in Hebrew [אֱנוֹשׁ] means 'man,'[bb] as in the first man. He also places the birth of Seth at 98,260 years before the Deluge.[19]

[bb] Strong's Concordance No. 583. Seth [שֵׁת] *Sheth* means *a son of Adam*, no. 8352. Adam [אָדָם] means mankind (no. 120) but comes from the root *Adom* [אָדֹם] which means *to be red*, no, 119.

GENESIS CHAPTER FIVE

The Genealogy of Adam and the Age of the Demigods

Adam lived 930 years, Seth begot Enosh and lived 912 years. Enosh begot Cainan and lived 905 years. Cainan begot Mahalalel and lived 910 years. Mahalalel begot Jared and lived 895 years. Jared begot Enoch and lived 962 years. Enoch begot Methuselah, walked with god, and was taken by God when he was 365 years old. – Genesis 5:1-24 (summarized)

The large timeline following the Anunnaki placed the creation of man about 270,000 years before the Deluge. At approximately 200,000 years before the Deluge, a new glacial period began, and life began regressing.[4] As theorized by Sitchin, Adam was a Neanderthal and lived 112,360 years, Seth, a Cro Magnon, lived 104,560 years, and Enosh, a *Homo sapiens sapiens*, lived 98,260 years before the Deluge.[19] This places the birth of Enoch at about 100,000 years before the Deluge. Chapters seven and fourteen dove into the *Book of Enoch*, and the character of Enoch is fascinating. The *Book of Jubilees* states that Jared was born during the days the Watchers came to earth to instruct the children of men. His father named him because of the times. The Hebrew rendering of Jared is *Hrd,* meaning 'to descend.' When Enoch was born,

"He was the first among men that are born on earth who learned writing and knowledge and wisdom and who wrote down the signs of heaven according to the order of their months in a book, that men might know the seasons of the years according to the order of their separate months." – Jubilees 4:17

In the *Book of Enoch*, Enoch reveals to his son Methuselah that he traveled to South America on his grand adventures with Shamash and

Adad. He went to the end of the earth and then saw a land that burned day and night:

"I saw at the end of the earth the firmament of the heaven above. And I proceeded and saw a place which burns day and night, where there are seven mountains of magnificent stones, three towards the east, and three towards the south. And as for those towards the east, was of colored stone, and one of pearl, one of jacinth, and those towards the south of red stone. But the middle one reached to heaven like the throne of God, of alabaster, and the summit of the throne was sapphire. And I saw a flaming fire." – Enoch 18:6-9

The Andes mountains, where Machu Picchu lies, consist of a chain of mountain ranges. In the north, there are three mountain ranges (Cordillera: Occidental, Central, and Oriental. Oriental runs parallel to the coast at an elevation of 3,730m) that run the span of the Andes with a plateau in the central range known as Lake Titicaca. The southern Andes are covered with glaciers (sapphire) and snow (pearl). The Aconcagua is the highest peak and the color of the jacinth stone. Ojos del Salado is the world's highest active volcano (flaming fire) and the second-highest mountain. It would make sense that Enoch would call a volcano 'flaming fire,' for it was not something that he would have seen before in Mesopotamia. Mount Fitz Roy in Patagonia is the color of red stone and is usually covered in snow. In the middle lies the mountain Illimani (The Shining One).[cc]

The *Sumerian Kings List* notes that Enmeduranki ruled in Sippar for 21,000 years, while the Bible says that Enoch only lived for 365 years. Sitchin assists with this confusion and notes that Enoch was born 75,040 years ago. He multiplies Enoch's age (365) by the sexagesimal Anunnaki count (60), making Enoch 21,600 years old when he was

[cc] Authors Note: At this point, I wish I could take credit for coming up with all the times "The Shining One" appears (Mountainiq.com).

taken.[20] The *Sumerian King List* writes the names of the first ten rulers before the Deluge. The timelines and story align with those told by Manetho before the Deluge when the demigods ruled after the gods.[21]

Enoch to Noah

Methuselah begot Lamech and lived 969 years. Lamech begot Noah and lived 777 years. Noah was 500 years old when he begot Shem, Ham, and Japheth. – Genesis 5:25-32 (summarized)

GENESIS CHAPTER SIX

Man multiplied, and the Nephilim came down and mated with the daughters of man. The Lord became upset and punished mankind, shortening their lifespans to 120 years. Giants existed and daughters of men bore children to giants. These giants were called the LU.GAL [Sumerian term meaning Mighty men]. – Genesis 6:1-4 (summarized)

The Lord saw a lot of wickedness on the earth and that all the people's intents, thoughts, and hearts were evil. The Lord regretted that He had made man and the animals, so a plan was devised to destroy everything He created from the earth. Noah found grace in the Lord's eyes. He was perfect in his generation and walked with God. The earth was corrupt before God and filled with violence. – Genesis 5:5-11 (summarized)

Enmeduranki[dd] was the King of Sippar, Lamech was the King of Shuruppak, and Noah was the last King of Shuruppak before the Deluge. While the Bible math calculates 869 years between the birth of Methuselah and the flood, Sitchin gives a larger time range of 71,140

[dd] EN.ME.DUR.AN.KI – whose name means "Ruler Whose ME Connect Heaven to Earth (Sitchin. *12th Planet*, p. 136.).

years.[20] Within this time frame, many earth changes were happening that are not mentioned in the Bible. The Sumerian stories also show that LU.GAL was also the name given to the intermediary demigod kings.

Noah's birth is found in the *Book of Enoch*, and it may surprise the reader to discover what made his birth so different.

"[Lamech's wife] became pregnant by him and bore a son. And his body was white as snow and red as the blooming of a rose, and the hair of his head and his long locks were white as wool, and his eyes were beautiful. And when he opened his eyes, he lighted up the whole house like the sun, and the whole house was very bright." – Enoch 106:2

This description sounds incredibly similar to all the descriptions of the Watchers and Elohim. The beings that radiated light were known as the Shining Ones. What is even more interesting is that after Noah was born,

"He opened his mouth and conversed with the Lord of Righteousness." – Enoch 106:3

If Lamech were the King of Shuruppak, the city of Ninmah's medical center, Noah would have been born in a birthing center because he was a king. Ea would have probably been present because (as other texts indicate) Ea was the father. When Noah exited the womb, he looked like Ea and spoke to him immediately. Lamech was overwhelmingly aware of what the Elohim or Anunnaki looked like that he felt and went to his father, Methuselah. However, it seems that Lamech could have also been a demigod because he was in the lineage of the *Sumerian King List* (Ubartutu). When he arrives at his father's home, he exclaims:

. . .

"I have begotten a strange son, diverse from and unlike man, and resembling the sons of the God of heaven; and his nature is different, and he is not like us, and his eyes are as the rays of the sun, and his countenance is glorious. And it seems to me that he is not sprung from me but from the angels, and I fear that in his days a wonder may be wrought on the earth." – Enoch 106:5-7

It was not because Noah may not have been the biological child of Lamech, but rather, his birth was the realization of warnings sent by Enoch about an upcoming cataclysm. During the lives of these demigod kings and biblical patriarchs, a lot was happening. Where the Bible notes that the flood happened because the Lord was mad at mankind, geological history tells a different story. There were three crustal displacements over 130,000 years before the Deluge (130,000 BCE, 40-50,000 BCE, 30-40,000 BCE).[22]

The Wisconsin Glacial period began approximately 50,000 years before the Deluge. The North American ice cap advanced three times, with ice core samples showing climate warming and cooling changes at 72,500 BCE, 24,300 BCE, and 11,800 BCE. Hapgood notes that the North Pole was located in Alaska at this time, yet within less than two thousand years, there was a rapid decline, and the North American ice sheet vanished.[23]

Geological data shows a large amount of time before the flood when massive amounts of earth changes caused the flood. Approximately 49,000 years before the Deluge, there was a complete magnetic field reversal with a pole shift of approximately fifteen to forty degrees, known as the "Laschamps Event." This magnetic pole reversal caused massive changes on the earth's surface. Hapgood remarks on the environmental changes due to large-scale volcanic eruptions after the crustal displacement.[24]

The spread of volcanic dust after an eruption causes a decrease in temperature, an increase in ice sheet growth, a decrease in food production and supply, and an uncommonly high increase in the spread of disease. There is also a high increase in rainfall, which causes excessive flooding. The drastic temperature fluctuations between climactic zones cause multiday violent gale-force winds and earthquakes.[24]

These changes are also referred to in the *Atra-Hasis* but are seen as Enlil's revenge towards mankind because they became too noisy. It tells how Enlil sent the šuruppu-disease to quiet mankind. Atrahasis appealed to Enki for the suffering of his people and received advice to stop the disease. The people became too noisy again, and Enlil called for the gods to cut off the food supply, stop the rain, decrease water flow to the rivers, and send strong winds. After drought and starvation came the second wave of disease. Man resorted to savagery, reproduction stopped, and seven years later, Enlil sent the flood (25,200 Earth years).[25] While it is unknown what caused the Deluge, Sitchin suggests that the Antarctic ice sheet was unstable. When Nibiru passed between Jupiter and Mars,[ee] the gravitational pull caused it to slide and caused the worldwide tidal wave.[26] At 11,600 BCE, the ice sheet severely declined, and the flood occurred.[4]

Making the Ark

The Lord came to Noah and told him about the violence on the earth and his plan to destroy everything through the flood. The Lord told him to make an ark of gopherwood and cover it with pitch. The Lord would bring the flood, and everything on earth would die. The Lord would then establish a covenant with Noah, and Noah and his family would enter the ark with two of every animal and all the food needed. – Genesis 6:13-22 (summarized)

. . .

[ee] The ADAPA beasts still served faithfully, but ENLIL had growing discontent towards them. He knew the AR (planet death ship) would be arriving soon and that it would cause a gravitational pull on the Earth. He used this knowledge to his advantage and plotted a course for the ship to have the greatest impact. The ship would land near the pole, causing glaciers to fall into the ocean and leading to extreme environmental changes. The planet had been cooling for an extended amount of time, reducing the amount of workable space on the planet. If the ADAPA drowned, oh well, the environmental changes were needed to warm the planet. ENLIL would give the ADAPA shelter, but the ADAMUS would be destroyed (Morningstar p. 36).

The Anunnaki made an agreement to let mankind die and not to tell any of them. Enki, however, did not like this plan because he cared for his creation.[ff] He went to Atrahasis and told him through a reed stick to tear down his house and make a boat with multiple decks and a roof like the Abzu (showing that Atrahasis/Noah was aware of Enki's home) and to cover the boat inside and out with bitumen.[27]

Utnapishtim,[gg] the son of Ubara-Tutu, lived in Shuruppak at the temple and heard the voice of Ea through the temple's reed wall telling him about the upcoming disaster and to build a boat. He told Atrahasis to tell the people that he was building a boat to leave the city and could no longer live in a place that worshipped Enlil while he worshipped Ea. The town came to help him construct the boat, then filled it with food and the *seeds* of all living things.[hh] The boat had six decks and nine compartments.[28] The flood story Berossus tells also states that Atrahasis took all the seeds of life with him and all his male and female servants.[29]

GENESIS CHAPTERS SEVEN AND EIGHT

The Flood

The Lord told Noah to go into the ark and told him the various types of animals to take with him. Noah was told it would begin to rain in seven days and last for 40 days and 40 nights. Noah was six hundred years, two months, and seventeen days old when the floodwaters came from the great deep, and the windows of heaven opened.

. . .

[ff] Ea took a group of ADAMUS beings into the underground caverns of the HEN-T hybrids (similar to the stories of the Hopi and Grand Canyon helpers and the underground caverns of Derinkuyu). Others were taken to the mountain highlands (similar to the Incan stories of Tampu-Tocco), and he created a special cargo ship to sail far out into the ocean and survive the disaster (Morningstar 36).

[gg] Atrahasis in Old Babylonia, Utnapishtim in the *Epic of Gilgamesh*, Ziusudra in Sumer, and Noah in the Bible.

[hh] There were not two of every animal but the DNA of all living things on the Earth (Sitchin p. *Divine Encounters*, p. 93-94).

The animals joined them on the same day that Noah, his wife, his son, and his son's wives entered the ark. All the flesh that contained the breath of life was in the ark. Then, the Lord shut them in.

The flood was on the earth for forty days, and the waters increased high above the earth as the ark floated on the surface. The high hills and everything on the earth were covered in water. The water rose 15 cubits (22 feet), and the mountains were covered. Everything on the land died. The water continued for 150 days.

God[ii] remembered Noah and all the living things and made a wind blow for a 150 days, causing the waters to subside. The flooding and rain stopped, and the water receded. The ark rested on Mount Ararat after seven months, and three months later, the tops of the mountains were visible.

At the end of 40 days, Noah sent out a raven to see if the land was visible, but the raven continued to roam. He sent out a dove, and the dove returned. Noah waited seven days and sent out the dove again. It returned this time with an olive tree leaf. After waiting another seven days, Noah sent out the dove, but it did not return this time. – Genesis 7:1-8:14 (summarized)

Within the Genesis verses are three different stories of flooding: two different verses of the animals and the ark, one account of the extreme wind, and the reference to 40 days is mentioned twice. Either the Biblical editor did a bad job of telling the story, or something else is happening.

Texts show that it was not only Ea who helped Atrahasis but rather it

ii Transition back to Elohim.

was a group effort to help save the humans. Out of the Pantheon of Twelve, Iškur/Adad, Nanna/Sin, and his twin children Utu/Shamash and Inanna/Ishtar were born on Earth and possibly had an affinity towards humans. Texts also show that Ninmah joined Ea to help save the humans.[26] When the gods learned of the coming disaster, they all had the opportunity to leave and return to Nibiru, yet they decided to stay and wait out the disaster in orbit. Ea told Atrahasis to look for the sign of Shamash to know when to close the door of the boat and prepare for the flooding. Shamash was in charge of the Sippar spaceship, and it would be clear from watching the actions of the gods as they left to close the door.[30] As Noah was in his boat with his family, tumbling through the waves, the gods watched from above and wept. Ninmah held the gods accountable.[31]

After the Flood

After one year exactly, Noah was able to remove the covering of the ark. A month and 27 days later, the earth was dry. Then God told Noah to leave the ark, and everything that was on the ark followed. Noah built an altar to the Lord and burned sacrifices on the altar. – Genesis 8:15-20 (summarized)

"And the Lord smelled a soothing aroma. Then the Lord said in His heart, 'I will never again curse the ground for man's sake, although the imagination of man's heart is evil from his youth; nor will I again destroy every living thing as I have done." – Genesis 8:21 NKJV

When the Anunnaki returned to earth, they smelled Noah's offering and gathered in joy. Enlil was angry because he had specifically stated that no one was to be saved. Ea stood up to Enlil and blamed him for the harsh treatment of humans with the Deluge when any other means of "punishment" would have been sufficient. Humbled, Enlil took

Utnapishtim's hand and stated that he and his wife were mere mortals but were now like gods.[32]

In *Erra and Ishum,* the king of the god's remorse is heard when he tells Erra that after the flood, he looked at what remained and grieved. He saw the offspring that were living, and like a farmer, he could hold the extent of his seed in his hand. Instead of helping, he made a house, cleaned his robes, put his crown on, and reveled in his glory. To avoid the truth of what he did, he put all the people back to work in the Apsu. Yet he knew that all the people, his actions, and the flood watched as he retreated into his house. Then, as mentioned in chapter five, he asked where the Seven Sages of the Apsu were because only they could make his body holy again.[33]

Ea calmed Enlil with persuasion and told him that without mankind, the gods would be doing the work again, and they would not exist as gods without humans. Enlil relented:[jj]

. . .

[jj] When ENLIL discovered the betrayal from EA, he was full of rage. Both brothers had betrayed one another and committed crimes against the ASA-RRR and the DAK Empire. They agreed that a truce was needed, or they would lose everything. ENLIL promised never to interfere with EA's development of life on ERIDU. EA promised that he or any other Anunnaki (K-D) would never interfere with ENLIL's command. In a truce, ENLIL gave EA the new implements for the beast, tools to help the beast learn to grow their own food. EA promised to increase the skills and abilities of ENLIL's hybrids used in ERIDU's administration. With genetic manipulation, ADAPA had minimal communication, understanding, and analytic thought. ADAPA could make small-scale decisions but still under servitude. ERIDU flourished, and peace was obtained for a time. The ADAMUS grew in number due to their uncontrolled procreation activities and made it difficult to produce specific traits via genetic manipulation. EA's EA-SU teachers would help grow the ADAMUS and give them a chance to understand beauty, appreciate art, and give them a sense of self-awareness. Without the influence of the ASA-RRR system, they would be able to have a sense of belonging and the power to choose their own destiny. Prince EA became a master Genesis scientist and became known as 'Lord (EL) of the Beasts' (EL-EA: LEO), and ENLIL became 'Lord of Obedient Servants.' Word spread around to neighboring star systems and galaxies about the ADAMUS. Another group of Genesis Scientist Lords arrived and gave EA a filament DNA strand to encode passion into the ADAMUS. Known as the AKHU, this group was a bipedal bird humanoid. They gave ADAMUS an invisible motivational force, passion, and deep emotions. This gift was called "The Gift of the Feather." As ERIDU grew, the ADAPA were obedient servants, and ADAMUS labored for food and supplies (Morningstar 36-39).

God blessed Noah and his family. God also placed everything on earth under the hand of Noah and his family. God made a covenant with Noah and all his descendants never to flood the earth again with the intent to destroy it. The sign of the covenant is the rainbow. When it rains and a rainbow is seen, it is sent as a reminder of the covenant between Noah, God, and all flesh on earth. – Genesis 9:1-17 (summarized)

The Descendants of Noah

"Now the sons of Noah who went out of the ark were Shem, Ham, and Japheth. And Ham was the father of Canaan. These were the sons of Noah, and from these the whole earth was populated." – Genesis 9:18-19 NKJV

The Curse of Canaan and The First Pyramid War

"And Noah began to be a farmer, and he planted a vineyard. Then he drank of the wine and was drunk, and became uncovered in his tent. And Ham, the father of Canaan, saw the nakedness of his father, and told his two brothers outside. But Shem and Japheth took a garment, laid it on both their shoulders, and went backward and covered the nakedness of their father. Their faces were turned away, and they did not see their father's nakedness.

"So Noah awoke from his wine, and knew what his younger son had done to him. Then he said: 'Cursed be Canaan; a servant of servants he shall be to his brethren.' And he said: 'Blessed be the Lord the God of Shem, and may Canaan be his servant. May God enlarge Japheth, and may he dwell in the tents of Shem; and may Canaan be his servant.' And Noah lived after the flood three hundred and fifty years. So all the days of Noah were nine hundred and fifty years; and he died." – Genesis 9:18-29 NKJV

This "curse of Canaan" has never made sense. Why would a son be punished for walking in on his father while he was asleep, even if he was drunk and naked? Ham just told his brothers; he did not boast about it to everyone. And why was Canaan punished for his father's act instead of the other children that Ham had? Either the punishment is overly severe, or the story is skewed.

After the deluge, approximately 10,500 BCE, Enki and Enlil divided the land and people into three regions: Mesopotamia and the surrounding regions (Shem), the Nile Valley down into Africa (Ham), and the highlands of Asia Minor around the Caspian and Black Sea (Japheth).[34] Noah's sons would go off in each direction and populate the region. A hierarchical sexagesimal order was already established upon the arrival of Enki and Enlil. City-states existed before the flood and demigods ruled as intermediaries between the people and gods. After the deluge, rebuilding and reestablishing each region was the first order—the gods were the literal land-lords over the regions and the people were the inhabitants.[kk] Figure 31 shows the relation of the Gods and their names across cultures.

- AN.U (60) – ruled in heaven with official wife AN.TU, his residence was in Nibiru.
- EN.LIL (50) – ruled over the airspace with official wife NIN.LIL (45), his residence was in Nippur.
- EN.KI (40) – ruled over the land and sea with his wife NIN.KI (35), his residence was Eridu.
- NIN.MAH (5) – ruled over and lived in Dilmun.
- NIN.UR.TA (50) "Lord Who Completes The Foundation" –

[kk] Each Anunnaki's epithet is listed in Table 9. The expanded family tree and Pantheon of the Anunnaki are shown in Figure 31. Sitchin notes that the pantheon of the gods in the Vedas are called Aryans or noblemen (Sitchin. *The 12th Planet*, p. 62).

ruled over Lagash with his wife BA.U and lived within the region at Girsu.

- NAN.NAR (Sin) (30) "Bright One" – ruled over and lived in Ur with his wife NIN.GAL (25).
- UTU (Shamash) (20) "The Shining One" – ruled over and lived in Sippar with his wife Aya. Son of Nannar and twin of Inanna.
- IN.ANNA (Ishtar) (15) "Anu's lady" – Not originally part of the Pantheon or a Great God. Daughter of Nannar and twin of Utu.
- ISH.KUR (Adad) (10) "Mountainous, Far Mountain Land" (10) – ruled over Sumer and Akkad with his wife Shala (she is rarely spoken about in texts).
- MAR.DUK "Son of the Pure Mound" – ruled over and lived in Egypt with his wife Sarpaint (rarely spoken about in texts).
- DUM.UZI "Son Who is Life" – Married to Inanna.

- NER.GAL "Great Lord of the Lower World," – ruled over and lived in the Lower World "Antarctica" with his wife Ereshkigal.[35]

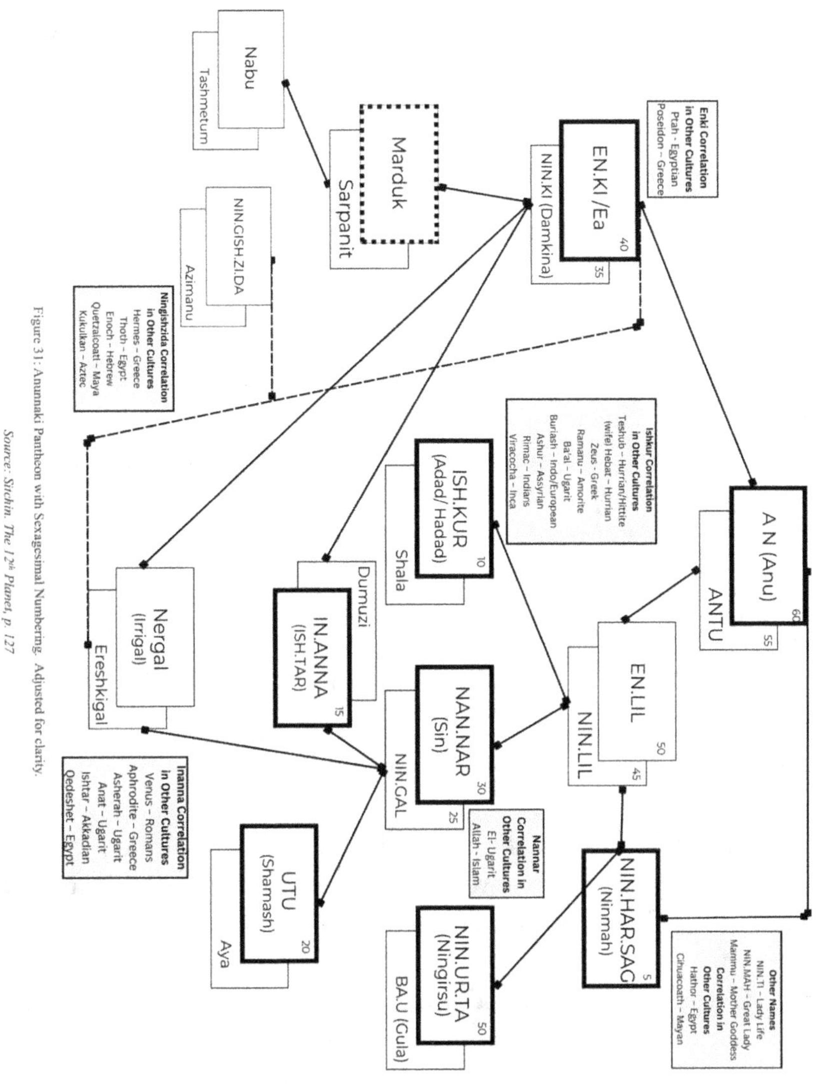

Figure 31: Anunnaki Pantheon with Sexagesimal Numbering. Adjusted for clarity.

Source: Sitchin. The 12th Planet, p. 127

The *Book of Jubilees* details the portions of land allotted to each brother:

The generations of Shem received the whole land of Eden, the land from the Red Sea to the land of India, the land of Mesopotamia, and the mountains of Ararat. The generations of Ham received the second portion of all the land south and to the right of the Garden, to the mountains of fire, and down to the sea on all sides. The generations of Japheth received the third portion north of the river Tinâ, as far as the sea waters. The land of Japheth is cold, the land of Ham is hot, and the land of Shem is neither hot nor cold but a blend of both. – Jubilees 8:20-30 (summarized)

The "Sin" of Canaan is explained three chapters later:

"Ham and his sons went into the land which he was to occupy, which he acquitted as his portion in the land of the south. And Canaan saw the land of Lebanon (Shem's land) to the river of Egypt that it was very good. He went not into the land of his inheritance to the west (that is to) the sea, and he dwelt in the land of Lebanon, eastward and westward of the border of Jordan and from the border of the sea.

"Ham, his father, and Cush and Miziram, his brothers, said unto him: 'You have settled in a land which is not yours, and which did not fall to us by lot: do not do so; for if you do, you and your sons will fall in the land and (be) accursed through sedition; for by sedition you have settled, and by sedition will your children fall, and you shall be rooted out forever... Cursed are you and cursed shall you be beyond all the sons of Noah, but the curse by which we bound ourselves to an oath in the presence of the holy judge, and in the presence of Noah our father.'

"But he did not hearken unto them and dwelt in the land of Lebanon from Hamath to the entering of Egypt, he and his sons until this day." – Jubilees 11:28-33

The specific land that Canaan settled in meant one thing only: a younger Enkite god used the people to gain access to the space sites under the control of the Enlilites. These sites included the Landing

Place at Baalbek, the Spaceport at Sinai, and the Mission Control Center of Giza. Manetho notes that Ra/Marduk divided Egypt into Upper and Lower dominions to correct the problem of succession and two sons, Osiris and Set.[36]

In 9780 BCE, the First Pyramid War took place.[4] Egyptian stories noted this battle in *The Battles of Horus* when Set attempted to usurp the two crowns of Egypt for himself as he moved from Upper Egypt to Lower Egypt.[37] Attempts were made for peace. Still, Horus and Set ultimately fought in the Sinai region, marking the first time the gods fought alongside men.[ll] For Horus' victory, the gods gave him the whole of Egypt. Set was banished to Asiatic lands in the east, becoming Enšag, the lord of Dilmun.[mm] The younger Enkite god attempted to extend his domain into the lands of Shem and was victorious for a time.[38]

The Second Pyramid War

Marduk wasted no time after his victory in the First Pyramid War and seized his moment to establish himself forever as a dual deity, Ra/Marduk. He changed his epithet in Egypt from the god MAR.DUK (Son of the Pure Mound) to IM.KUR.GAR RA (Ra Who Beside the Mountainland Abides). His epithet also changed in Babylon from KA.DINGIR "Gateway of the Gods" to KA.DINGIR.RA (Ra's Gateway of the Gods).[39] This ambition to become the supreme ruler of the Earth caused Ra/Marduk to stop at nothing.

[ll] The ASA-RRR way was always war. The children of EA and ENLIL were constantly at war with one another to achieve power. As long as the Empire was not jeopardized, wars were seen as a way for the younger generation to learn the ways of the Empire. With the enhancements given to the beasts, they were used by the younger gods for administration in their individual kingdoms and to control the other beasts. EA's worst nightmare had been realized, but he understood that free will was the most important part of his experiment. The beasts learned war and became pawns in the hands of the Royal's chess matches called war. The Royals had access to reanimation chambers and could be brought back to life or healed from the verge of death. The beasts, however, were expendable (Morningstar 39).

[mm] The story of the birth of Enšag is found in *Enki and Ninmah: A Paradise Myth*.

The Second Pyramid War was between the younger gods in 8970 BCE.[4] The *Lugal-e ud me-lám-bi nir-ǧál* text describes the all-out war to get Marduk out of Egypt. Manetho notes that Horus was only in Egypt for 300 years after Set.[21] Ninurta led the frontal attack on his ship called IM.DU.GUD (Divine Storm Bird) with weapons while Adad snuck behind enemy lines and destroyed food supplies. The enemies retreated into the mountains of the Lower World. Meanwhile, Ninurta and Adad covered the mountains with fire. They used the Stormer to rain poison down on the cities, flatten the land, dye the skies wool red, and fill the rivers with the blood of the innocent bystanders.[40]

Ninutra claimed victory, received the title of Vanquisher, and celebrated his swift victory in the Lower World. Yet this victory was in vain.[nn] Azag (Marduk),[oo] gave strict orders to his brothers to show no resistance to the attacks and to make it appear as though they won. The weapons of the other gods were hidden and covered in the earth. Azag sent word to the sons of Enki to retreat into the Great Pyramid. Enki and Thoth then raised an impenetrable protective shield around the Pyramid. While Ninurta was celebrating in the Lower World, Nergal snuck out during the night with his ship and weapons. He arrived at the Great Pyramid and helped strengthen the defenses of Enki's side.[41]

When Ninutra discovered their deceit, he was outraged and had Shamash cut off the water supply to the Pyramid from the Nile. The Enkites barricaded themselves in the Pyramid while Ninurta waited to

---

[nn] Victory speech of Ninurta is found in *Mythen von dem Gotta Ninib* (Myths of the God Ninurta).

[oo] Sitchin notes ASAR is part of an epithet name (Sitchin. *The Cosmic Code*, p. 196). The Seventh Tablet of the *Enuma Elish* lists Marduk's 50 epithets, three from the main epithet of Tutu, meaning to make anew or bring restoration. (1) King's translation of *Enuma Elish* notes Tutu Azag: Zi-azag – The Bringer of Purification, Stephany's translation of the *Enuma Elish* notes this epithet as ZIKU – who insists on cleanliness and giver of abundance. (2) Tutu Aga-azag – The Lord of Pure Incantation and quickener of the dead, Stephany translates AGAKU – Lord of the immaculate spell who brings resurrection and compassion to the captives. (3) Tutu Mu-azag – Pure Incantation destroying the evil ones, Stephany translates TUKU – flawless incantation that eradicates all malevolence with spells. Sitchin also notes A.ZAG was a derogatory epithet given to Marduk by Ninutra (Sitchin. *The Wars of the Gods*, p. 222; King. L. W. *Enuma Elish*, p. 131-33; Stephany. *Enuma Elish*, p. 46-7).

attack. Horus disguised himself as a ram and attempted to sneak out but was hit by the Brilliant Weapon and blinded.[pp] Ninmah (Hathor) cried out to save her son's life. She had reached her limit and put her foot down. Her children had taken over her home, cut off her water, and were once again raging war against each other. She stepped out of the Pyramid, walked across battle lines, and called out for her son Ninurta.[qq,41]

Negotiations for Peace

Ninmah was met with care by Ninurta and discussed the way towards negotiations. The Enkites (Enki, Marduk, and Nergal) exited the pyramid and met with Enlil, Ninurta, Sin, and Adad at her residence. Enlil and Enki agreed to each other's requests.

- Enki asked for:
    - He and his descendants would have sovereignty over the Giza complex for all time.
    - He would regain his access to Eridu and reopen Edin as a holy house.
    - His lineage would be allowed to enter and travel through Mesopotamia freely and help bring prosperity back to Mesopotamia.
- Enlil asked for:
    - The land of the Shem would rightfully be restored to Enlil's lineage, including:
        - The Sinai Peninsula that housed the Restricted Zone.
        - The Radiant Place that housed the Mission Control Center and would one day become Jerusalem.

---

pp This story is found in Egyptian *Legend of the Ram*. The Ram god of Egypt is Khnum/ Khnemu/ Khoum.

qq Sitchin notes Ninmah's actions were written about all over Mesopotamia in a text called "I Sing the Song of the Mother of the Gods" (*Wars of Gods and Men*, p. 165). It is also found under the title "Temple Hymn 7: The Kesh Temple of Ninhursag," "Temple Hymn 42: The Eresh Temple of Nisaba," and "A Hymn to Nisaba."

    - The people of Ham and Canaan would leave the land of Shem and return to their territories.
- Other items on the agenda:
    - The Great Pyramid would be destroyed and never used again as a weapon.
    - Marduk was not allowed to enter Egypt or Mesopotamia.
    - Ningishzida would be placed as Lord over Giza and Lower Egypt.[42]

The Destruction of the Great Pyramid and its Prisoner

In 8670 BCE, Ninurta fulfilled his part of the peace negotiations and walked into the Great Pyramid for the first time. He walked into the heart of the pyramid (the Queen's Chamber) and destroyed the stone of destiny that was the "heartbeat of the pyramid." He then traveled to the hidden chamber (King's Chamber), removed the housing net that "surveyed heaven and earth" from the stone box, and destroyed the UGU stone. He destroyed the rope that hermetically sealed the three portcullises as he left the secret chamber. While he walked down the Grand Gallery, he either destroyed or repurposed the 27 stones that lined the floor and caused the Gallery to glow the color of the rainbow. After Ninurta was finished inside the pyramid, he made his way to the apex stone and let it crash onto the ground. Ninurta made it known that the Great Pyramid was to be made into a mountain of stone for the benefit of all future generations. [43]

The Short-Lived Time of Peace

After the end of the Two Pyramid Wars and the exile of Marduk from Egypt, Ningishzida ruled over Egypt as Thoth, the Great Pyramid was destroyed, and the Age of Peace finally arrived in time to bring in the

Age of Cancer in 8650 BCE (Figure 32). It was the age when demigods ruled over the land as the gods focused on rehabilitation efforts. Before the Deluge and the two Pyramid Wars, the gods had only focused on gold production. After all the disasters and death, the gods focused instead on the future of their operations with mankind.

Before the gods returned to rebuild their pre-established cities, they complained to Enki that he had been withholding the Divine Formulas of Civilization. They asked for the formulas to be shared so that all of Sumer could prosper. After the olden cities were established, Enlil reaffirmed Ninurta's position as second and was again placed in charge of the Olden Land of Lagash. Adad's lands were extended to include Lebanon and the region of greater Canaan. Nannar's territory was expanded to include modern-day Syria and the Sinai Peninsula. Lastly, Enlil appointed Shamash as the ruler and commander over the City of SHU.LIM, 'The Supreme Place of the Four Regions' (Jerusalem).[44]

To extend his new region, Nannar/Sin established the city of Jericho circa 8500-7000 BCE and the lost city of Tell Ghassul circa 7500 BCE. Enki's sharing of the Divine Formulas progressed civilization into the Neolithic Period of 7400 BCE. Before this era, mankind was not allowed in the Olden Cities. Now, mankind was allowed into the cities to assist with service to the gods, including gardening, cooking, artisans, priests, and musicians.[45] This caused the olden cities to transform into urban areas, causing the development of the King's Highway connecting Africa to Mesopotamia and Arabia.[44] The lands of the Anunnaki flourished as peace reigned; the Age of Gemini in 6480 BCE turned into the Age of Taurus in 4320 BCE.

Around 4000 BCE, the gods constructed E.ANNA (House of Anu) to prepare for the visit of Anu and Antu in 3800 BCE. Along with celebrations, the gods came together to discuss the future. The god's agenda is shown in the Akkadian version of *The Tamarisk, the Palm, and the King*. Anu, Enlil, Enki, Shamash, and Ninmah are listed with the other gods. Together, they decided to establish new cities and give kingship to mankind. The king would be an intermediary between the gods and men. Together, the gods chose Gušur to rule over the black-headed people in Kish because they all loved him.[46] Kish was placed under the control and care of Ninurta, and it would become the first administrative capital of Sumer in 3760 BCE. The kings would be appointed by Enlil and would be called LU.GAL (Mighty Men). The city of Ur was to be established as the economic heart of Sumer for Nannar/Sin.[45]

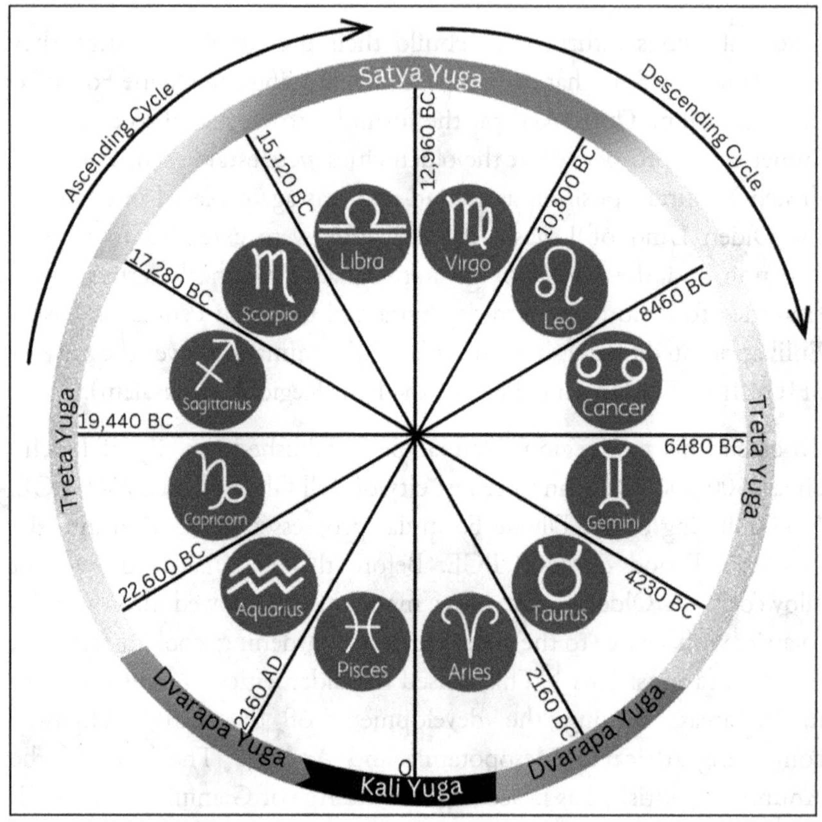

Figure 32: Precessional Cycle of the Great Year with Yuga Cycle

Genesis Chapter Ten

Known as the Table of Nations, the genealogy of Noah is given. Japheth's lineage is given, and they formed the coastland people of the Gentiles according to their language, families, and nations. The lineage of Ham is repeated from the previous chapter. – Genesis 10:1-7 (summarized)

. . .

"Cush fathered Nimrod; he was the first on earth to be a mighty man. He was a mighty hunter before the Lord; therefore it is said, 'Like Nimrod a mighty hunter before the Lord.' The beginning of his kingdom was Babel, Erech, Accad, and Calneh, in the land of Shinar. From that land he went to Assyria and built Nineveh, Rehoboth-Ir, Calah, and Resen between Nineveh and Calah (that is the principal city)." – Genesis 10:8-12 ESV

Canaan's lineage is listed, and nine family groups are established. They spread all throughout Canaan from Gerar to Gaza and settled around the Dead Sea cities of Sodom, Gomorrah, Admah, and Zeboiim, according to their language, families, and nations. – Genesis 10:15-20 (summarized)

The verses in Genesis 10:8-10 tell the Akkadian and Babylonian story of when the gods handed kingship to man. The Hebrew spelling is used for both Kish (Cush) and Ninurta (Nimrod). The *Sumerian King List* shows that when kingship was lowered from heaven,[rr] it was in Kish,[ss]

[rr] The ADAMAS beasts had proven themselves through the guidance of EA and the other ANUNNAKI. In time, the beasts were allowed to run small kingdoms of territory as long as they continued to pledge themselves to the Royals and serve the ASA-RRR Empire. The ADAMAS beasts had risen from slave laborers to productive members of society and above all the other creatures on ERIDU, yet still strictly second to the LORDS. Wars of the royals continued as kingdoms rose and fell at the expense of the ADAMAS. ERIDU entered into the era of MARDUK's takeover. He was an accomplished DAK warrior and held the title of Master. Wanting kingship over ERIDU, he challenged his father, EA, his uncle, ENLIL, and other family members for the throne. He would not let his father EA's deprivation of the throne become his fate (Morningstar 39-40).

[ss] Kingship in Kish lasted 24,510 years per the *Sumerian King List* and does not match up with the storyline of Sitchin. Multiple kings listed are demigods who ruled Kish, Uruk, and Ur, such as Etanna, Dumuzi, and Gilgamesh. These kings may have reigned during the time of the demigods and the biblical story increases confusion due to intermingling of events and no concrete timeline of the biblical narrative.

and Gušur reigned for 1200 years.[46] While Genesis asserts that Babel (Babylon) was the next to gain kingship, the *Sumerian King List* shows kingship moved to Erech, then Ur (Sumer).[47] There is a large discrepancy that is pointing towards something larger.

GENESIS CHAPTER ELEVEN

"Now the whole earth had one language and one speech. And it came to pass, as they journeyed from the east, that they found a plain in the land of Shinar, and they dwelt there. Then they said to one another, "Come, let us make bricks and bake them thoroughly." They had brick for stone, and they had asphalt for mortar. And they said, "Come, let us build ourselves a city, and a tower whose top is in the heavens; let us make a name for ourselves, lest we be scattered abroad over the face of the whole earth.

"But the Lord came down to see the city and the tower which the sons of men had built. And the Lord said, 'Indeed the people are one and they all have one language, and this is what they begin to do; now nothing that they propose to do will be withheld from them. Come, let Us go down and there confuse their language, that they may not understand one another's speech.' So the Lord scattered them abroad from there over the face of all the earth, and they ceased building the city. Therefore its name is called Babel, because there the Lord confused the language of all the earth; and from there the Lord scattered them abroad over the face of all the earth." – Genesis 11:1-9 NKJV

After a long absence, Marduk arrived on the scene in 3460 BCE and wanted kingship for himself.[4] An Akkadian tablet, *The Legend of the Tower of Babel*,[tt] tells the story of Marduk's arrival into the region and his actions as the instigator of building the temple. The text does not

---

[tt] William Bramley suggests that the Towel of Babel event was intentionally caused by the Custodians. He defines the custodians as an extraterrestrial society that has ownership of

state the name of the instigator because of the extensive damage to this tablet, but all the clues point towards Marduk. His heart was evil towards the father of the gods and corrupted the people around him to achieve his purpose. Enlil arrived on the scene to discover Marduk was the one who instigated the entire debacle. Enlil went to his mother, Damkina, who stood at her son's side. As the text reveals, once again, Marduk's numerical rank among the gods is the problem. When Marduk refuses to stop building the tower, Enlil ends the attempt at night and confuses the language.[48]

While there is debate about the location of the Tower of Babel, Sitchin speculates that the Genesis narrative gives away the location with the description "tower whose top is in the heavens." A ziggurat matching this description is found in Babylon known as the seven-stage pyramid, named E.SAG.ILA (House Whose Head is Lofty). This ziggurat was to be a new Gateway of the Gods, but Marduk's attempt to create his own Spaceport failed, and he was banished from all of Mesopotamia. He returned to Egypt to return to his former glory as Ra. When he arrived in Heliopolis, everything had changed.[49] Instead of vying for supremacy, Thoth left Egypt, took a group of Kushite followers from Nubia, arrived in Mesoamerica in 3113 BCE, and became the Olmec and Toltec god Quetzalcoatl.[50] Marduk's brother Nergal was in the Lower World of Antarctica, and Gibil inhabited the Abzu. Dumuzi, however, increased his domain to border Upper Egypt and, with the help of his wife Inanna, he would become a hindrance to Marduk's supremacy.[49]

After the tower incident, Marduk was on shaky ground with the other gods. The story turns to Inanna and Dumuzi, who stood by each other's side even during the battles between the other gods. Almost all the gods supported their union when the gods were not battling each other. Marduk, however, was adamantly against their union.

After their union, Inanna and Dumuzi could not produce an heir and conspired for Dumuzi to lay with his younger sister, Geshtinanna.

---

the Earth and its inhabitants since before the time of antiquity (Bramley. *Gods of Eden*, p. 38.).

Dumuzi attempted to talk to Geshtinanna about his intentions, but she refused. Dumuzi then raped his younger sister. Marduk heard and was furious. Dumuzi was remorseful but knew he would be reprimanded like Enlil before he married Ninlil. He was called to the council of the gods and taken by Shamash, who let him go. He was captured and escaped two times more with the help of the Enlilites, and while trying to get across a river, he drowned. Not knowing if it was accidental or intentional, Inanna was furious and called for Marduk to be placed on trial.[uu] *Inanna and Ebih* tells the story of Marduk hiding in the pyramid as she attacks and calls for his death. Enlil arrives and decides that due to circumstances surrounding Dumuzi's death, Marduk would be buried alive and imprisoned in the Great Pyramid, and it would be sealed closed.[vv] Egyptian stories note when Ra was absent from them, and they called him Amen-Ra (The Hidden One Ra).[ww,49]

After Thoth left for Mesoamerica, Dumuzi's death, and Marduk's

---

uu The story of Geshtinanna and Dumuzi is found in "A Song of Inana and Dumuzi." The stories of Dumuzi's death are found in "Dumuzi's Dream," and "The Death of Dumuzi," by Samuel Noah Kramer. The story of the trial is found in "Myths of Inanna and Bilulu" (Bilulu, another name for Marduk), "The Most Bitter Cry," and "Inanna and Ebin."

vv *The Terra Papers* echoes that Prince MARDUK was wrongfully accused of the death of DUMUZZI. He took refuge in the pyramid, as a council was called, and decided to seal him in the pyramid to die. Luckily, MARDUK had faithful allies in the SSA-TA rebels hidden underground. They helped him escape through a tunnel beneath a pyramid, and he fled to the heavens. MARDUK met with the SSA-TA reptilian rebel group and promised great power, full participation within ERIDU, and wealth in exchange for support in overthrowing the ASA-RRR empire on ERIDU and taking his place as ruler. The SSA-TA rebels were happy to oblige as enemies of both AN-U and the Queens of the SSS-T Throne. They saw an opportunity to weaken the DAK Empire and the Queens of SSS-T and grow their own rebellion within the ARI-AN Empire. The SSA-TA warriors agreed, and within a short time, they increased their forces exponentially. Their ultimate goal was to control the Ninth Passageway regardless of whether or not MARDUK was successful, or a failed civil war broke out. The SSA-TA waited for the moment to strike (Morningstar 40-41).

ww The date at which the wedding of Inanna and Dumuzi, death of Dumuzi, and imprisonment are confusing in the text. *The Wars of the Gods*, p. 232, notes that 3450–3100 BCE was a chaotic time all caused by Marduk. The wedding, death, Tower of Babel, imprisonment, and exile of Marduk all occurred within this time frame. Marduk was evicted from Egypt after the Second Pyramid War and from Babylon after the Tower of Babel incident.

imprisonment, the gods also gave kingship to mankind in Egypt. In 3100 BCE, the First Pharaonic Dynasty began and Mizraim, the son of Ham, became the first pharaoh. Egyptian records note the name Menes. Waddell remarks in the commentary that Hebrew *Mizraim* transliterates to *Misri* in Egyptian. He is also called Mineus by the Africanus scribes.[51] In the Ethiopian King list, E.A. Wallis Budge notes the first dynasty after the flood and the Tower of Babel was called Kam, which is related to the Egyptian origins of the name. The king list preceding Kam was removed because they were regarded as serpent worshippers (Enkites were descendants of the serpent family). The people of Ethiopia, known as Abyssinians, claim that they are descendants of Ham by way of his son Kush.[52] The Kushite or Nubian kingdom is regularly mentioned in Egyptian history, steles, and the El Amarna letters. The 25th Egyptian Dynasty was called the Kushite Dynasty.

The death of Dumuzi resulted in what Sitchin calls the age of Ishtar. A new Third Region was created in the Indus Valley and placed under her dominion. Kingship was transferred from Kish to E-Anna, with the first king listed as Meskiaggasher, son of Utu, who initially had dominion over E-Anna. His son Enmerkar worked with Inanna and transformed E-Anna into Uruk. Inanna ruled over Aratta and worked with the two lands to help them flourish. Inanna wanted to become more powerful and help her lands flourish, so she obtained seven of the MEs from Enki.[53] Sitchin suggests that both Mohenjo-Daro and Harappa were the cities of Inanna, and Harappa was known to the Sumerians as Aratta.[54] Inanna was placed in the Pantheon of the Twelve Great Gods and replaced Ninmah.[55] During the Age of Inanna, the people thought the gods had left, causing chaos throughout the region.

The long timeline places Inanna's dominion over Uruk at 2900 BCE.[4] On the other side of the Nile River, the Brotherhood of the Snake was formed under Pharaoh Khufu.[xx,56] In 2371 BCE, Inanna fell in love with Sargon I, and the Akkad Empire was established. Sargon I

---

xx *The Terra Papers* notes the HEN-T that were part of the coup were called SHET-I and were strategically placed within Egypt to rule as pharaoh and serve as administrators and loyal servants of MARDUK (Morningstar 45).

attempted to expand into Babylon, offending Marduk, and then fell from grace. Inanna was still furious with Marduk after his return to Babylon. Nergal then teamed up with Inanna. *Erra and Ishum* describe how he betrayed his brother and destroyed his house, which had devastating consequences for the surrounding region. Marduk went into hiding.

Inanna had increased her power and started to step on other gods' territory by having the new king, Naram-Sin of Akkad (called Agade in Sumerian texts), attempt a coup against all the other gods. The gods were finished and decided to end Inanna's reign and Akkad. *The Curse of Agade* details the story of their battles and the ultimate death of Naram-Sin, the fall of Akkad in 2260 BCE, and notes that Akkad was never to be restored.[57]

In 2160 BCE, Amon departs from Egypt, marking the transition from the Old Kingdom Dynasties to the Intermediate Period. In 2123 BCE, Ur-Nammu became King of Ur under Nannar/Sin. The 3rd Dynasty of Ur began, and Sumer flourished like never before. Edifices of the great gods were restored throughout the region, including agriculture, and they showed the promise of such a bright new beginning that Enlil and Ninlil returned to the city. As a mighty warrior of Nannar, he brought peace to the region using the Divine Weapon and then died on the battlefield.[58]

In 2095 BCE, Shugli succeeded Ur-Nammu, and Sitchin notes it was the same year that Abram and his family arrived in Ur. Shugli continued in his father's footsteps by enhancing the empire of Ur. In 2055 BCE, he created a treaty with the Philistines and became King of the Four Regions. Yet in 2048 BCE, he died and was succeeded by his son Amar-Sin.[59] After all the major political changes in Ur, Abram and his family began their journey towards Canaan but stayed in Haran for 24 years.[yy]

[yy] MARDUK began silently planning his coup. He seduced the HENT-I servants and administrators of ENLIL by using every manipulation possible, even promising them second only to him. The underground HENT-I approached the servants of ENLIL and various royal families. Once the SHET-I (the secret ones) was established, they began recruiting ENLIL's command forces, logistics and communications forces, and all of the

In 2048 BCE, Marduk arrived in Harran after hiding in Hittite lands and worked with his son Nabu to build a following of Marduk in Borsippa. Abram was told by Yahweh to leave Harran and journeyed to Canaan. He arrived in the Negev and then went to Egypt to obtain military reinforcements from the pharaoh. Abram returned in 2042 BCE with reinforcements for the battle and found the Canaanites had converted to worshipping Nabu.[60]

Marduk was determined to cause the Age of the Ram to come even though it was before time. When Abram arrived in the Negev, he intended to gain the Spaceport for Yahweh and defend it. After Marduk successfully obtained the spaceport, he returned to Babylon as their god. Nergal and Ninurta's lands were in constant battle with the Babylonions who were desperately trying to stop the age of the Ram. Nergal had finally had enough and decided to end Nabu, have Marduk bury him, and then kill Marduk. In *Erra and Ishum*, Ninurta (Ishum) attempted to stop and reason with Nergal, but it was useless. They agreed on targets: Spaceport at Mount Most Supreme and Kings Highway, then warned the other Anunnaki of the pending events. The next day, they set out with the Awesome Seven Weapons, and as Lot was leaving Sodom, his wife turned around and was turned into a pillar of salt.[61]

The Lamentation Over the Destruction of Sumer and Ur details the aftermath of the Awesome Weapons and the evil wind that followed the nuclear fallout. The gods fled for their lives as the people perished.[62] The Age of the Ram finally arrived. Marduk had defeated the other gods and was granted kingship, as Babylon was untouched by the Evil Wind. The gods assembled and recognized Marduk's NAM.TAR (destiny that cannot be cut). They all agreed and proclaimed, "Marduk was King,"[zz]

---

departments within ENLIL's administration. Even the DAK warriors stationed in ships to assist the ANNUNAKI with uprisings had switched and became loyal followers of MARDUK (Morningstar 41-42).

[zz] The DAK warriors above ERIDU, sent other ships on detours. The paradise planet turned into an island outpost controlled by MARDUK and the SHET-I. ENLIL and his followers returned to the ASA-RRR Empire. EA took many of the ADAMUS back with him to his star system Pleiades (Place of BAAL [Lord] EA DA [The Creator]), and the

and the Tablets of Destiny were handed down from Enlil. With the tablets, Enlil handed the divine bow weapon to Marduk and transferred his numerical rank of 50 to Marduk. The fifty titles of Marduk were to be remembered forever in the *Enuma Elish*. Marduk had successfully absorbed the Pantheon of the Twelve and was one god.[63]

Hammurabi became King of Babylon in 1800 BCE, which caused Babylon to flourish. By 1760 BCE, Marduk had created a distinct star religion, and Nibiru had been forever changed to Marduk in celestial worship.[4] In 1660 BCE, the Kassite Dynasty in Babylon began, and their secret ties to the Egyptian Brotherhood of the Snake were found in El Amarna letters.[64] While the Kassite Dynasty in Babylon possibly began before Thutmose III or Akhenaten's time (depending on variations within Egyptian history), it ended in 1160 BCE, well after the construction of El Amarna ended. When Akhenaten built his new capital at El Amarna (Akhenaton), he included a large temple for the Brotherhood shaped like a cross. This was a specific move by the Brotherhood to move towards an outwardly monastic lifestyle. In reality, it was to appear as if the church and state were separate while secretly controlling both. When sending correspondence between members, it was common to refer to each other as "brothers."[65]

In the introduction of *The El Amarna Letters*, Moran comments on the brotherhood as an alliance between the various rulers, and the first four Amarna letters were exchanged between the pharaoh and the Kassite king, Kadashman-Enlil I, who ruled from circa 1374-1360 BCE, with four references to the brotherhood and each calling one another brother.[66] Kadashman-Enlil I also introduced the Kassite Cross into the world and was depicted with kings across the region from Assyria to Egypt.

---

other ANNUNAKI offspring were forced to flee. MARDUK ordered the search of ANNUNAKI and their heirs, still on ERIDU. They were given the choice of complete enslavement or death. MARDUK then destroyed all records of other ANNUMAKI by altering the stone monuments, edifices, obelisks, and tablets. MARDUK was now the LORD GOD and CREATOR OF THE UNIVERSE, and he was named the "Sun God RAA." His records were retroactively changed, and to ensure his supremacy, the ADAMA and ADAPA colonists were lured into chambers and promised wealth and power but were instead lured into brainwashing and mental reprogramming. If colonists refused, they were forcibly taken for adjustment (Morningstar 42-43).

Over time, it was transformed into the Rosicrucian cross, and Ashurnasirpal II was depicted wearing it in a stele his son made after the capture of the Phoenician coastlines.[67]

THE AGE OF THE PROPHETS

In the 8th century BCE, the biblical prophets began speaking of the destruction of Jerusalem as the return of Nibiru and the Age of Pisces drew near. It was known across the region that the great gods had already left. Only Nannar/Sin and Marduk were left, as the other gods had traveled across the seas. Fighting between the two sides (Enkite vs Enlilite) continued as none of the previous gods that assisted man were present, and man was truly on his own.[68]

Amos, Hosea, Isaiah, Jeremiah, Ezekiel, and Malachi all spoke of destruction to come before the Day of the Lord arrived. The Assyrian and Babylonian kings prepared for Marduk to return. In 744 BCE, Tiglath-Pileser III, King of Assyria, launched an attack, causing the partial exile of Israel. His son and successor, Shalmaneser V, conquered Israel in 722 BCE, and the Ten Tribes of Israel disappeared from history. In 612 BCE, after the death of Ashurbanipal, Medes attacked Assyria, and in two years, the great empire was gone.[69]

King Nebuchadnezzar of Neo-Babylon restored Babylon to its former glory and rebuilt the seven-stage ziggurat. Sitchin notes that this ziggurat was successfully transformed into the new DUR.AN.KI, and to ensure that the transfer from Jerusalem to Neo-Babylon was complete, he captured Jerusalem in 598 BCE. He fulfilled the prophecy in 2 Kings 23:27-27 and Jeremiah 21:7. Nebuchadnezzar believed that he could defile the temple because the general belief was that Yahweh had left, just as the other gods had. In 562 BCE, Yahweh punished Nebuchadnezzar for his actions against Judah, Jerusalem, and the Temple. He died in agony, as prophesized in Jeremiah 50:18. Sitchin notes that this was the year that Nibiru passed by Earth, and the gods left to return to Nibiru.[aaa,70]

[aaa] Before the EA-SU were taken by force, they fled into the wilderness or mountains to

Yet the Age of the Ram had not drawn to an end yet. In 556 BCE, Nabonidus (noted as Nabu-Na'id by Sitchin) was chosen by Sin in Harran to rule over Babylon. Yet, to rule over Babylon, a King needed to be chosen by Marduk. As chance would have it, that same year, the prophecy of the Day of the Lord arrived as an abnormal solar eclipse over the Mesopotamian region. Sitchin notes the cause of this "abnormal" solar eclipse was the passing of Nibiru over the region.[71] Nabonidus stated that the appearance of Nibiru/Marduk marked his divine appointment as king to rule over Babylon. In 555 BCE, he ascended the throne of Babylon, yet began restoring the temple of Sin in Harran and Babylon. The *Nabonidus Chronicle* details the ninth year of his reign, 546 BCE.

In 538 BCE, Cyrus the Great, King of Persia, walked into Babylon, held the hand of Marduk's statue, and stated that he was called by Marduk to become ruler of all. He was crowned King of Babylon, ended the exile of Israelites, and told them to rebuild the Temple in Jerusalem. To the Israelites, he was the one Yahweh called to end their exile and restore the Temple. This act united Marduk and Yahweh. After he died in 529 BCE, his son Cambyses marched to Pelusium, defeated the Egyptians, and proclaimed himself pharaoh in 525 BCE. Aware of Egyptian custom, he stated that Ahura-Mazda chose him as pharaoh.[72] Ahura-Mazda is also known as Ormuzd in the Persian Mithras Mystery Teachings.[73] He bowed in reverence to the Egyptian gods, and the priests

---

write their knowledge in stone. They knew they would forget, but one day, the truth would be uncovered, and the clues would be found. Towers were built to send electric signals around the globe and dull the Gift of the Feather within them. This electric signal assisted with blanketing the planet from outside contact. Houses of Obedience were created, and the beasts were obligated to attend obedience lessons every seventh day. The SHET-I turned the truth of the past into fables and myths that rational men had no time for. They twisted history and turned EA into "the Evil One." The Golden Era of ERIDU, Prince ENLIL, Prince EA, and the SIRIAN lords were all gone. RAA knew that the SHET-I could not be trusted and placed his children in the position of RA-KA Pharaohs. One night, the SHET-I quietly took control of everything. They went to capture RAA but found that he escaped. They forced RAA's offspring and royal court to undergo the mental erasure the ADAMA and ADAPA experienced. They would be faithful to the SHET-I reptiles only. Yet the beasts knew that there was more, even if they could not remember quite what it was (Morningstar 43-46).

granted him the title "Offspring of Ra."[72] Within 13 years, Amen-Ra, Marduk, Yahweh, and Mithras became one god and forever changed history. In 500 BCE, the Age of the Fish arrived.

UNCOVERING YAHWEH

Who was Yahweh then? The ancient stories of extraterrestrial contact preceded the Bible by thousands of years. There are many names for God throughout the Old Testament. Modern translations use variations of 'God' to cover these names. Still, the truth is found in the footnotes of some versions, as shown in Table 9.[bbb] Other names, such as Adonai, have also been substituted to reference god.

Verse	Name	No.	Hebrew	Translation
Genesis 1:1	God	430	אֱלֹהִים	*Elohim*
Genesis 2:4	LORD God			*Yahweh Elohim*
Genesis 4:1	LORD	3068	לַיהוָה	*Yahweh*
Genesis 5:22	God	430	הָאֱלֹהִים	*Ha Elohim*
Genesis 14:18	God Most High	410, 5945	עֶלְיוֹן אֵל	*El*[ccc] *Elyon*
Genesis 17:1	God Almighty	7706	שַׁדַּי	*El Shaddai(y)*
Genesis 31:33	The Eternal	5769	עוֹלָם	*El Olam*

Table 9: Names of God in the Bible

O'Brien notes that just like the term Elohim, Yahweh is not a name but instead a title. In Exodus 3:14, when Moses asks the name of the one speaking to him through the bush, the Lord says to Moses, "I am that I am." The Hebrew rendering is "Ehyeh - Asher - Ehyeh." *Ehyeh* is the singular first-person form of *Elohim*. The Sumerian root of Asher is *ash*, meaning 'perfect' or 'one.' Thus, it should be translated as "Ehyeh one Ehyeh." O'Brien translates this as "Ehyeh, the first of the Elohim."[74]

Another version is given by Sitchin, who notes that it is not present

---

bbb Biblegateway.com (online access to obtain almost all printed versions of the Bible) was used with Strong's Concordance to obtain variations of names. Some versions do not contain footnotes for all forms.

tense but future tense and should be rendered the simplistic version of the one who was, is, and always will be. *El Olam* is consistent with this and referred to as the eternal, unchanging. While *Yahweh*'s characters are inconsistent with the unchanging part, *Olam* is mistranslated and presents Yahweh ineffectively. Instead of *Yahweh* saying that he is eternal, *El Olam* means that Yahweh is king and lord over many *Olam*'s or worlds.[75]

Continuing from O'Brien's translation, Yahweh could have told Moses that he was just one of the Eloha. The singular form of Yahweh agrees with all the ancient texts around the world. Every Anunnaki god is seen in the various characteristics of Yahweh.

- Enki
  - When Elohim hovers over the waters with his team of fifty.
  - When Elohim created man.
  - When Yahweh Elohim saved Ziusudra and all the other Noah characters.
  - When Yahweh Elohim gave knowledge and sciences to the people from Proverbs 2:6 and 30:4.
  - The Yahweh Elohim, also known as Nudimmud (He Who Fashions) that Psalm 119:73 speaks of.[76]
- Enlil
  - When Yahweh Elohim gets angry at Adam and Eve and sends them out of the garden.
  - When Yahweh Elohim gets angry at mankind and sends the flood.
  - The Master of Eden.
  - The Supreme Elohim or El Elyon, who is Lord of Heaven and Earth in Psalm 97:7.
  - The Yahweh Elohim of Psalm 82:1 and Psalm 29:1, who reigns as Commander in Chief over the Elohim.[76]
- Ninmah
  - When Yahweh Elohim regrets the destruction the flood caused and was sad.
  - The Yahweh Elohim that does not forget mankind in

Deuteronomy 4:19, 10:14, 1 Kings 8:27, 2 Chronicles 2:5, 6:18, and Ecclesiastes 12:2.[76]

- Shamash
  - When Yahweh Elohim is described as a judge.
  - When Yahweh Elohim gives laws to the Israelites.
  - When Yahweh Elohim is described as his face shining in Psalms.
  - When Yahweh Elohim walked with Enoch.
- Adad
  - When Yahweh Elohim walked with Enoch.
  - When Yahweh Elohim's voice was described as thunderous.
  - When Yahweh Elohim rides in the clouds in Psalm 104:1-3.
  - When Yahweh Elohim introduced themselves to Abram as El Shaddai.
    - Hebrew *Shaddai*, means 'omnipotent.'
    - Akkadian *Shaddu*, means 'mountain.'
    - ISH.KUR, 'god of mountains,' whose domain was over the Taurus Mountains in Asia Minor.[76]
- Ninurta
  - When Yahweh Elohim is the warrior hero of the Israelites in Numbers 21:14, Numbers 15, and Isaiah 13:4 and 42:13.[76]
- Nannar
  - The Yahweh Elohim, whose name (Sin) no one could explain.[77]
- Anu
  - When Yahweh Elohim ruled from the heavens.
- Marduk
  - When Yahweh Elohim interacts with Abram.
- Inanna
  - When Yahweh Elohim decided to take over the lands of the other Elohim.
- Nergal

- When Yahweh Elohim sends destruction upon Sodom and Gomorrah.

They are all Yahweh throughout the Old Testament. Yet when the Age of the Ram arrived, Marduk took on all the attributes of the great gods, according to *Enuma Elish* Tablets VI and VII. However, even Marduk and his son Nabu are ultimately defeated by Yahweh because all the gods left after the Age of the Ram ended. After Cyrus the Great and his son Cambyses, the major gods at the time, Marduk, Aten-Ra, Yahweh, and Ormuzd, the creator of Mithra, became one god, merging ancient solar worship with stories of the extraterrestrial ancestors. If the reader has been following the story in *The Terra Papers*, they will note that this is precisely what Bek'ti said. Those who were left after the great coup were the rulers. Only at the beginning of the Age of the Fish (Pisces) did Yahweh change into the Brotherhood of the Snake.

Part II Summary

Part II of this book was a constant cycle of learning one thing: seeing a crazy correlation that could not be ignored, diving deep into a rabbit hole, and then figuring out how to get out, knowing that the only way out of the rabbit hole was through. Growing up Christian and learning about all the twisting and turning of what the Church teaches as doctrine but is fiction is a large pill to swallow.

It was the Anunnaki who came to the Earth to mine for gold. They rehabilitated the planet after the deluge, got tired of working so hard, and created mankind. They fought wars, dealt with demi-gods and their shenanigans, launched nuclear missiles over rulerships, and in the end, the Egyptian sun god Ra/Marduk was victorious. He worked with Abram of the Bible to assist in his final battle. He then rewarded Abraham with a covenant. The pharaohs were secretly learning twisted information from mystery schools of the Brotherhood. Thutmose III and Amenhotep III worked with the Brotherhood to begin the transition toward monotheism and created the greatest dynasty in Egyptian history. The Brotherhood then moved away from the pharaoh's palace yet systematically began transforming from polytheism to monotheism through their chosen pharaoh, Akhenaten.

Akhenaten was renamed Moses, and he spoke with Yahweh in his UFO. Moses built an advanced communications device with holographic capabilities. He and the other thousands of people who Moses and Joshua led in Israel suffered the effects of radiation exposure, and most died. Solomon built a more sophisticated holographic device to transmit signals and communications to an orbiting craft above the Earth's atmosphere. Elijah was taken by these beings from the UFO, and Ezekiel told the most incredible description of the UFO possible. Josephus, Livy, and others told stories of ancient extraterrestrial contact and UFO sightings. The Anunnaki gods were aspects of Yahweh, and Yahweh was merely placed as a title instead of using their names. Yet, due to the political, social, and religious changes happening within the ancient world, the actual truth is difficult to discern because everything was altered, and the truth was muddled. When the Age of Pisces arrived in 500 BCE, the Great gods had left, and now it was time for the Church to become Yahweh.

Part Three
The Plot Twist of Biblical Proportions

How ancient history was intentionally restructured to create a narrative fit for the Church, created a society dependent upon the church, and lost its roots from the stars; and the Brotherhood's rise to dominance through secrecy and mankind's ultimate victory.

Discovering the Creation of Jesus and the Church – The Mysteries from Antiquity to Constantine the Great

It is imperative to understand today what led society and the Christian church away from the knowledge of humanity's ancient extraterrestrial ancestors. *The Terra Papers* end with the HEN-T reptilians usurping control from Marduk and taking over the planet after the Anunnaki gods left Earth (Edin). Zechariah Sitchin's stories align with Bek'ti and show that the gods left around 550 BCE when the Age of Aries changed to Pisces.

But what does the Age of Pisces have to do with the church and the Bible?

Absolutely everything! The symbolism alone is enough to write an entire book about, which, thankfully, Jordan Maxwell already did in *That Old-Time Religion*. He spends 17 pages explaining 55 points why the story of Jesus correlates to the sun and zodiacal understanding. First, it was common in antiquity to relate the Sun as a symbol of the Divine. Like the creator of all, the sun sustains everything on Earth; without it, there would be nothing.

- Hindus called the Sun Lord Brahma/Krishna.
- Depending on the Upper or Lower Kingdoms of Egypt, the Sun Lord was called Amun/Osiris/Horus/Ra.
- Chaldeans, Phoenicians, Hebrews, and Assyrians called the Sun Lord Bel/Adonaï(s)/Lord.
- Ethiopians called the Sun Lord Achōr.
- Persians called the Sun Lord Mithras.
- Grecians called the Sun Lord Apollo.[1]

Just a few of Maxwell's 55 points are needed to understand the Age of Pisces and the church:

1. Matthew 14:17 &19: God's *sun* feeds the five thousand with two fish and five loaves of bread.

- The "two fish" is the astrological symbol of Pisces.

2. John 14:2-3 ESV: "In my Father's house are many rooms...and I will come again and will take you to myself, that where I am you may be also."

- In the zodiac, each month is assigned a house. There are 12 *houses* in heaven. Each room has a different zodiac sign: Father's house – heaven, many rooms – zodiac house. The *sun* is in the House of Pisces.

3. Matthew 28:20: God's *sun* "is with you always." King James Version

uses the phrase "end of the world," while the English Standard Version uses "end of the age."

- The phrase "End of the World" is a common teaching the Brotherhood uses. Known as "Judgment Day" or "Doomsday Teachings," these are effective across every religion, race, tribe, and creed. It successfully removes the power of the incredible human being, places it on an unseen deity, leads to the idea of "Obey or Die," and, as successfully employed by John Calvin in the 1500s, shows that there is an unalterable plan or law that humanity can do nothing about.[2] The ESV wording is more appropriate because it states "Age," as the *sun* is in the house of Pisces until the end of the age. Chapter twenty noted that the end of the Age of Pisces is approximately 2026-2106 CE; thus, humanity is in the last days of the "Age."

4. Luke 22:10 ESV: "A man carrying a jar of water (Aquarius) will meet you. Follow him into the house that he enters."

- Follow the *sun's* path as it travels from Pisces to Aquarius.[3]

PRECESSION AND THE YUGAS

Precession

One of the great secrets of the universe is the language of numbers. Nikola Tesla said that the numbers three, six, and nine hold the keys to the universe. It is just one of the many incredible things he offered this world and an insight into esoteric teachings. These numbers help to understand the significance of The Great Year or the Precession of the Equinox. The earth's tilted axis causes the polar position to change and changes the constellations visible behind the sunrise during the vernal equinox. One complete cycle of Precession is 25,920 years.

- 12 – Number of constellations in the zodiac
- 30 – Number of degrees for each zodiacal constellation

- 72 – Number of years required to complete processional shift of 1 along the ecliptic
- 360 – Total number of degrees in the ecliptic
- 72 x 30 = 2160 – Number of years it takes for the sun to pass through one zodiac constellation
- 2160 x 12 = 25,920 – Number of years it takes for the sun to pass through all 12 zodiacal constellations

The Piscean Age Characteristics

The zodiacal Age of Pisces represents faith, religion, and sacrifice. This is easily identifiable with the correlation between Jesus and the. Each zodiacal age is 30 within the ecliptic, and within each Age are three 10 segments lasting 720 years. The three divisions are mutable, fixed, and cardinal. The Age of Pisces began in 500 BCE, and the changes can easily be seen in the segments and timetable of Figure 32.[a]

- 0-720 CE, the mutable decan – characterized by people needing direction and looking to follow, even at their own expense. This is the decan in which the Age is finding its identity.
- 721-1440 CE, the fixed decan – characterized by stagnation and an unwillingness to change. This is the decan that the Age is established and remains in this state throughout its duration.
- 1441-2160 CE, the cardinal decan – characterized by leadership, bold decision, and initiative action. This is the decan that the Age attempts to maintain its dominance with whatever means necessary before the change to the new Age.

---

[a] The dates of the Ages may be off by 100 years. Graham Hancock included the Mayan Calendar in his calculations and stated that the movement from Pisces to Aquarius was in 2012. Other theories note a lag time of 100 years at the end and beginning of each Age.

Yuga Cycle

Another type of time and Age measurement is the Vedic Yuga cycle. The yuga cycles represent the gradual decline and ascension of humanity's consciousness. One complete cycle consists of 12,000 years descending and 12,000 years ascending. This cycle is completed every 24,000 years. Within one cycle (yuga), four divisions signify types of changes that occur on the planet, including the changes in consciousness. Ten sets of 1200 years make up the yugas, with periods of mutation at the beginning (dawn) and end (dusk) of each yuga.

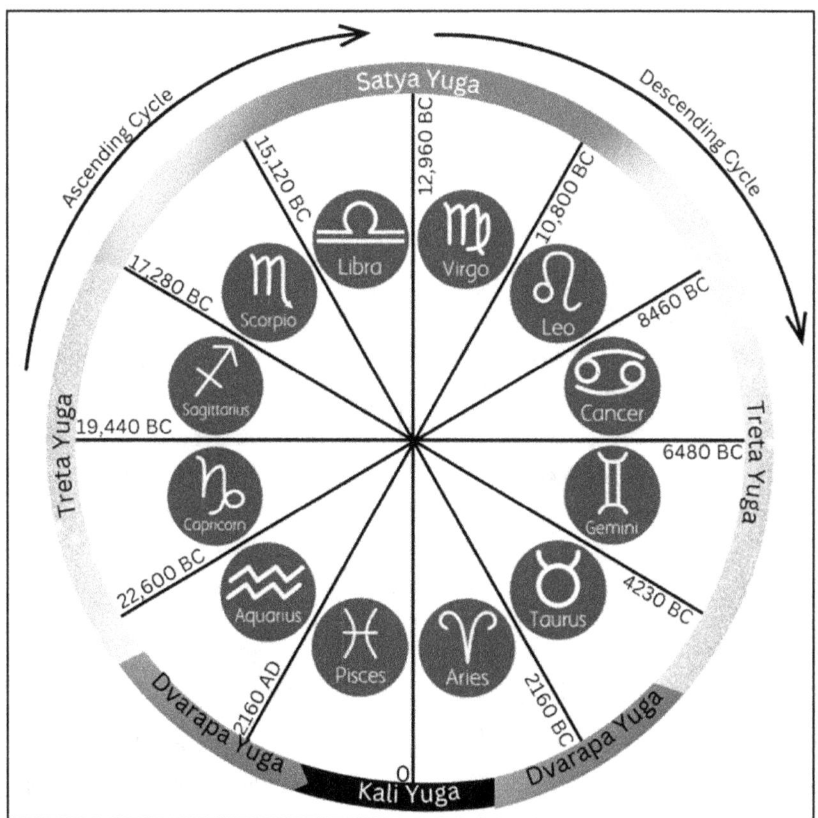

Figure 32: Precession of the Equinox and Yuga Cycle

Satya Yuga – The Golden Age (Spiritual Age) is 400 years (dawn) + 4000 years + 400 years (dusk).[4]

Tetra Yuga – The Silver Age (Mental Age) is 300 years (dawn) + 3000 years + 300 years (dusk).

Dvapara Yuga – The Bronze Age (Energy Age) is 200 years (dawn) + 2000 years + 200 years (dusk).

Kali Yuga – The Iron Age (Material Age) is 100 years (dawn) + 1000 years + 100 years (dusk). The Kali Yuga cycle began in 700 BCE and ended in 1400 CE.[b]

The cycle begins with 4800 years of the Golden Age of spirituality and consciousness and descends into the Silver Age for 3600 years. It continues to descend into the Bronze Age for 2400 years and finally into the Iron Age for 1200 years. The cycle then begins its ascension, coming out of the Iron Age to the Bronze, then Silver, and back to the next Golden Age. As the rest of this book outlines history to see how the Brotherhood acted out their grand plan, understanding the movement from yugas and through the three decans of the zodiacal age is important.

Secret Societies

Knowledge of spirituality, divinity, and the mysteries of the universe were once as knowable and understandable as science is today. Mystery schools taught the mysteries of mankind and higher esoteric knowledge.

---

[b] Swami Sri Yukteswar stated that in 1894, 194 years of the Dvarapa Yuga had passed, and in 1899, the dawn of Dvarapa would move into Dvarapa. This marks the end of the Kali Yuga in 1700 CE (Sri Yukteswar. *The Holy Science*, p. xvii-xviii).

They were open to anyone willing to embark on the quest towards liberation and the stabilization of the soul. Those in the ancient mystery schools knew that mankind was moving into the Kali Yuga, as did the Brotherhood. The ancient mystery schools prepared and placed their teachings in symbolism and allegory with the knowledge to unlock the mysteries when mankind would ascend to the Dvarapa Yuga.

The Brotherhood knew this, and just as the story shows in *The Terra Papers*, they hijacked the mystery schools for their twisted purposes and agendas. Other mystery schools were created to attempt to hold onto ancient wisdom. Yet as the ages moved toward the solidity of the yuga and into the fixed Piscean Age, the ancient esoteric knowledge was slowly lost. Mystery schools became secret societies, and these societies became the new "it thing." Societies were created for various other purposes without true esoteric wisdom.

Classification of these societies includes Religious (Egyptian Mystery teachings), Military (various Knights from the Crusades), Judicial (Vehmic Tribunal),[c] Scientific (Alchemy), Civil (Freemasonry), and Anti-Social (Garduña).[5] Specific divisions were crossed and lines were blurred over the millennia. Hall explains that all of these societies have attempted to create an enlightened society based on the definition of the times. Throughout the different eras of history, they all seemed to meld into one. In ancient times, they were mostly religiously and philosophically focused. In medieval times, they morphed into philosophical and political entities. Today, these societies are politically and socially focused while claiming to be religious yet void of actual spiritual truth.[6]

The greatest of time was taken during the creation of Part III and IV to create a comprehensive yet brief deep dive into how the cycle of time has, to an extent, dictated the events and circumstances.

The Rosicrucians

---

[c] The Vehmic Tribunal was a secret court and Holy Order during the Medieval Age that included the Ruler(s) and Free Judges. Free Judges included only pure-bread Germans and were to judge the actions of those who did not follow the Roman Church—specifically created to ensure that people did not revert to paganism after the Papal Inquisitions (McCall. *The Medieval Underworld*, p. 110-111).

When looking into the history of the Rosicrucian Society, there are many conflicting dates for its inception. Some sources even note that the Rosicrucians made up the age of this society and used terms such as 'fake antiquity.' H. Lewis Spencer in *Rosicrucian Questions and Answers* addresses this confusion. He notes that the varying dates are simply different revivals of the Order, and the confusion is tied to Freemasonry, whose origins are found in the Rosicrucian Order.[7]

As mentioned in chapter twenty, a secret Brotherhood originated in Egypt, with its first documented teaching dating back to the reign of Khufu in 2900 BCE. The teachings were practiced during the reign of Ahmose I, and all the kings of the 18th Dynasty were either teachers within the unnamed Brotherhood or masters. Thutmose III established the foundations of the Brotherhood, cloaked it in extreme secrecy, and created a cycle of 216 years. The first part of the cycle is 108 years of outer activity and recruitment, while the second cycle was to be a period of concealment and silence as though it did not exist. Each new active cycle of 108 years was a "new" Order without any connection to the preceding ones. Every separate branch within the Order had different dates for their cycle yet still followed the strict 216-year cycle.[8] Due to the active/inactive cycle, the Rosicrucian story becomes blurry and interweaves with other stories, but still operates today.

The only definitive string showing the origin of this Brotherhood—which was cloaked in extreme secrecy—was how kings and pharaohs addressed each other in their correspondence.[9] The Amarna letters between kings at the time of Akhenaten show they constantly refer to themselves (the kings and the pharaoh) as brothers.[10] Lewis uses these letters to show the actual antiquity and notes references to the Illuminati (the Illuminated Ones), the highest degree within the Brotherhood.[8]

Research into the Rosicrucians becomes murky and muddled because sources give different information on the purpose of the Rosicrucians, mix it with Christianity, describe their involvement in other movements, and that they constantly shift from good to evil. One thing is for sure: the secrecy and veil remain today.

. . .

The Foundation of the Cover-Up

At the end of the last Golden Age, mankind was aware of their divinity and lived in peace and harmony. As the cycle descended, the consciousness of man descended. Spirituality began to decline, and mankind believed that the gods were outside themselves, creating myths that represented the old knowledge—more time passed until the next age arrived. Mankind began to see the symbols of the myth as the things to be worshipped and understood. Then came a time when mankind served an invisible god, unaware of the divine, and reduced everything to the material.

There was only one divinity that existed within and animated everything. These were the original teachings found in the ancient traditions. Now, they have become defiled and perverted shells of the original divinity: a hollow substitute that men called God.

As Empires arose in the ancient world, rulers exerted their ideals and robbed humanity of their innate truth. The origin of secret mystery teachings was born. Bek'ti's story of the Annunaki shows that Ea, Thoth, and the Seven Sages were the ones who went around the world spreading this ancient knowledge. Other writers argue that the motives for control were the same because they were part of the group. Truth, however, can always be found in the origin of teachings, whether good or bad.

The Invasion of the Aryans

The origins of the Aryans are unknown.[d] They arrived in India during the 18th Egyptian Dynasty (1600 BCE), and with their arrival came modern Hindu beliefs. Chapter twenty detailed the activity of the Brotherhood during the reign of Thutmose III and Akhenaten. Once at

---

[d] Morningstar notes that the ARI-ANs (ARI – master, AN – heaven) are a race of ruthless reptilian war masters who used manipulation and wars to gain control over the Ninth Sector of the ERIDANUS galaxy. The SSS Queens were incredible politicians who specialized in war tactics. When the SSS Queens encountered the remaining victims of conquered regions, they employed mind control and manipulation techniques to create slaves for their reptilian empire. (Morningstar. *The Terra Papers*, p. 4)

El Amarna, the Brotherhood branched out and began enacting their plan of total and complete domination. India became a major Brotherhood network as they worked to transform their first society into an obedient feudal caste system via religion.[11]

Brahmin today has over 300,000 gods that create caste systems and keep the people in a life of servitude and division. Another hallmark of the Brotherhood's teaching is division and endless servitude, common to all Asiatic traditions today. The perverted Magi (priests) taught people they were not capable of understanding the Spirit as they shrouded the truth in figures and incomprehensible language. The Gymnosophists, or "The Magi of Brahminism," were the most severe upholders of the law. Along the banks of the Nile, they lived almost naked and revived many Asiatic religious philosophies teaching one God and the soul's immortality. They were close with the priestly colleges of Egypt and Ethiopia and met yearly across the Nile to offer sacrifices to Amon and celebrate the festival "Table of the Sun."[12]

While Brahminism became perverted in India, others emerged over the ages to combat its teachings. The *Manusmṛti*, also known as *Mānava-Dharmaśāstra* (The Laws of Manu), predates the birth of Moses but reads just as Genesis 1. Buddhism preached equality of all men yet was exterminated in India in 240 BCE.[8] Over time, Hinduism blended the *Vedas*, the teachings of Buddha and Mohammed, and the Aryan religion.[13]

Zoroastrianism

The origins of Zoroastrianism contain as much mystery as the *Rig Veda* and predate the arrival of the Aryans. Modern Zoroastrian teachings (Mazdeism) appeared around the time of the Aryans. Modern Zoroastrianism appeared around the same time frame when Kyouxares (grandfather of Cyrus the Great) founded the Median Empire.[13] The teachings of modern Zoroastrianism contain elements seen throughout all religions.

Zoroastrian teachings state: the eternal light created the first emanation, the chief god Ahura Mazda (Ormuzd), and is the lord or spirit of knowledge and wisdom. He made six Genii (messengers) and 28 Izads (spirits) to watch over the celestial world that existed for 3,000 years. The highest of the Genii and Izads was Mithras, Lord of the Sun and intermediary between men and the gods. Mankind was created, and Ormuzd benevolently ruled mankind for 3,000 years. Abrimanes was the second emanation of the eternal light, equal to Ormuzd, but grew jealous of his power and became the Evil Spirit that struggled for the soul of man for 3,000 years. At the end of the Fourth Phase of 3,000 years, Ormuzd will destroy Abrimanes, who will bow to him on the throne.[14]

Egyptian Mysteries

The ancient Egyptian mysteries were perverted just as much as all the other mystery teachings. The perversion started within the priest caste, who taught the exoteric doctrines of Osiris and Serapis while following the esoteric doctrines of Isis within the priesthood. Whether the priesthood did this to hide the truth from those who wished to pervert the original teachings is unknown. Within the esoteric mysteries, it was the privilege of the highest initiates of the priesthood to vote in the election of new kings.[15]

Symbols such as cosmology were used as an outward expression of the internal powers of the Eternal within each human. The symbols of the sun, moon, and stars began to take on a life of their own and were worshipped under new ideals of seasons and dependence upon them. The symbol then moved from the divine creator to the king. As the gradual decline continued from age to age, pure nature-wisdom became mythology and then romance. It changed with the regions, customs, and society in which the tradition traveled.[16]

Grecian and Pagan Mysteries

Pythagoras was an initiate of the Greek and Egyptian mystery schools. He used the knowledge he received from the mystery schools and the scientific teachings of Greece to express the mysteries through numbers and geometry. While society descended more towards the material, Pythagoras used geometry to express the ancient mysteries of antiquity and continue the transmission of truth. Today, they are known as the Eleusinian Mysteries, found in the Essenes and Therapeutae traditions, who studied the ancient mysteries in Alexandria around the time of Christ.[17] H. Spencer Lewis notes that the Essenes and Therapeutae formed during an active cycle of the Rosicrucians.[18]

Manicheism

Known as "son of a widow" in Masonic Lodges today, Manes was a slave sold to a widow in the 3rd century CE. He became a philosopher and attempted to establish a universal religion reconciling Christianity, Buddhism, Zoroastrianism, and Greek philosophy. With all these incredible teachings, the one teaching that stained all religions equally was his creation of the Devil. He rejected Judaism and taught that Jehovah was evil. In 277 CE, he was crucified by the Persian Magi. He created the Manichean Bible, and those who followed his teachings were the most despised sects by the Catholic Church.[19]

The invention of the Devil led to the Black Mass of the Middle Ages and created the rituals of black magic. Miller remarks on the dogma and rituals found in Manicheism and their similarities to Freemasonry and the Templars, including degrees within the society and signs. Yet the practice of Manicheism was only allowed to the adepts of Brotherhood Freemasonry.[19]

The followers of Manicheism were known as lawless anarchists and terrorists. This society employed their weapon of secret murders frequently. Manes created his following with the true purpose of systematically attacking the church and state to destroy society from within and reduce it to chaos.[19] From Manicheism came the Cathares (South France), Albigenses (North Italy), and Waldenses, who were persecuted throughout Europe. The Manicheans were renamed the Bohemian Brethren in England.[20] The Waldenses remain a distinct sect of Protestants in many regions of Europe.[21]

CYRUS AND ALEXANDER THE GREAT

- **600-530 BCE** – Cyrus the Great, Emperor of Persia
- **356-323 BCE** – Alexander the Great, Emperor of Macedon

Chapter twenty tells the story of when Cyrus the Great conquered Babylon, and his son Cambyses ruled over Egypt. Their successive and concurrent rule over vast regions marked the beginning of merging ancient beliefs. While it started with merging Yahweh, Amen-Ra, and Marduk, it was also the first time Ahura Mazda emerged.

Cyrus and Alexander changed the political landscape for a large population by successfully taking over empires. Alexander the Great claimed to be a descendant of Jupiter (Ammon). The myths say that when he was in Egypt, he met with Amen-Ra, who called him his son. As other stories have shown, myths usually have a way of being correct. He may not have visually seen Ra (because Marduk had already left Earth), but the priests would have ordained him as a son of Ra, similar to the story of Cambyses. This "Son of God" could have been a way of forming the story into the people's minds that a ruler is *chosen* by the supreme deity to rule and continue the Egyptian tradition of pharaohs.

Like those before him, Alexander the Great came from a specific lineage. His mother descended from the royal blood of Epirus, whose lineage dated back to Zeus. He was educated by Aristotle at the Temple of the Nymphs in Mieza, where the Cult of Mithras held teachings before it was closed by Constantine.

Cyrus the Great and Alexander the Great were followers of Mithras from Zoroastrianism. However, history specifies that they followed Mithras as their deity instead of identifying them as followers of Zoroastrianism. On the surface, it seems minute, yet it shows the movement towards the worship of a singular deity who was the intermediary between good and evil forces. As the Essenes waited for a new avatar of

Zoroaster to arrive in Palestine, the Judeans awaited the arrival of their savior to save them from the oppression of the Roman empire.[19]

Creation of the Christian Church and Its Values

The Mithraic Cult and Sol Invictus

After the introduction of Zoroastrianism at the beginning of the Persian Empire, it was degraded throughout the centuries. The merging of cultures following the conquests of empires led to the blending of the Apis cult, the Eleusinian Mysteries, and an increased focus on war and battle. Zoroastrian teachings were lost, and Mithras seemed to have formed his own religion. The exoteric teachings were incredibly popular among Roman Legion soldiers. Worship quickly spread throughout Rome as the esoteric Cult of Mithras grew. Nero was an initiate within the cult, while the official government policy was against established religions within the capital.[1]

218-222 CE – Marcus Aurelius Antoninus (posthumously Elagabus) was a Roman Emperor who attempted to make Sol Invictus the national god of Rome. He was assassinated for his attempt.

270-275 CE – Aurelian was the Roman Emperor.

. . .

December 25, 274 – Sol Invictus became the official religion of Rome with the first celebration of *Dies Natalis Sol Invicti*, a festival day of the Invincible Sun on the Winter Solstice.

Despite the limited information available about the Mithraic Cult, the fragment of the catechism, the Mithras liturgy account, and the depictions found in reliefs and sculptures provide intriguing glimpses. The consistent narrative depicted in the reliefs is that of Mithras slaying the Bull (Apis) while gazing up at the Sun (Sol Invictus) to the left with its crown of rays, and above Mithras is Luna with the crescent moon. The creation of this narrative during the time of the Biblical Christ's life adds an element of mystery. In Zoroastrianism, Mithras was the name of the sun and the intermediary between Ormuzd and Abrimanes. In Mithraic beliefs, Mithras became a man.

Figure 35: Mithras Relief – Vatican Museum

Credit: Gaius Cornelius CC: BY-SA 3.0,

The Mithraic Cult was solar-focused, referred to as the emperor cult,[2] and had seven grades of initiation. Each grade was associated with planetary bodies, and Sol Invictus was the sixth-level initiation. All cult activities were similar to other secret societies and took place in caves.

With the various significant conquests of Cyrus the Great and Alexander the Great, they employed the tools of the Brotherhood. They kept their societies active in an endless cycle of unstoppable global change while simultaneously becoming the heroes who saved society. They also used another tool of the Brotherhood seen only in hindsight

—support and destruction of countries for the sole purpose of furthering the Brotherhood's agenda.

CONSTANTINE THE GREAT

During the transition between the descension and ascension of the Kali Yuga, when humanity was furthest away from its own divinity, Constantine saved the day. Constantine's dreams of power began when his father, Constantius, was Caesar of Britain and Gaul. When his father died, Constantine assumed his father's position but did not let the junior status of that role within the tetrarchy of the Roman Empire hold him back. He quickly moved to establish himself as the Caesar of the newly colonized highlands of the West Roman Empire. Hungry for ultimate rule, he used Emperor Diocletian's dislike of Christians as his ticket. While the national god of Rome was Sol Invictus, Constantine, with his strategic understanding of the political landscape, knew that many Christians in East Rome held high positions within the state administration and were persecuted.[3]

He proclaimed himself to be the protector of Christians to gain the audience and trust of those in administration, and they did not have to know that there were no persecuted Christians who lived in his region.[4] In 312 AD, Constantine looked to the clouds for guidance on his way into battle and saw the words "In this sign, you shall conquer," with the cross. With this sign, he knew that Christ was on his side and he would be victorious.

Constantine should have fired his minister of propaganda and hired a fiction writer to come up with a better lie because none of that happened. He did not need it to be true; he just needed others to believe it was true. A year later, Constantine issued the "Edict of Milan," establishing tolerance for all religions. In 321, *Dies Solis* became the day of the Sun and the Roman day of rest. In 325, he became the sole Emperor of the Roman Empire and declared Christianity the official religion of Rome. The first Council of Nicaea met.

To attend the momentous Council of Nicaea were the Christian leaders of Alexandria, Antioch, Athens, Jerusalem, Rome, and every pagan leader of the cults, including Bacchus, Jamus, Zeus, Oannes, Osiris, Isis, and Sol Invictus.[a] Historically, the "Most Important Item" of such meetings was to address the Aryan Controversy. However, the real agenda of this council was to "unify" the competing beliefs to create one universal Catholic Church and create the character of Jesus Christ.[4] Jesus' name came from the Druid's 'Hesus,' who was crucified with a lamb and an elephant (the sins of the world),[5] and 'Christ' was derived from the symbolism of the Egyptian word *karast,* meaning 'embalmed mummy who was to rise again.'[6] Constantine and the Council of Nicaea had successfully created the perfect avatar for the Age of Pisces.

Following the popularity of Mithras, all the Mithraic holy days became Christian holy days. Symbols of pagan religions, such as the universal symbolism of the egg,[b] were inserted into Christian holidays. After Jesus was created, the gods of the other cults became the twelve apostles or saints of the Roman Catholic Church. Canon law and the Nicene Creed were established.[7] Bishop Eunomius attempted to voice the fraud that had taken place in the council but was in turn charged with fraud by the Church.[5]

Constantine effectively brought the Aryan feudal and serfdom system of Brahminism into the European world. The Brotherhood plan was on track. Their attempt with Akhenaten in Egypt had failed, yet with Constantine and the Age of Pisces, the Brotherhood succeeded, and access to pure spiritual knowledge was blocked. The Brotherhood could elevate themselves while robbing the people of human decency.[8] H. Spencer Lewis notes that the "Great White Brotherhood,"[c] as he calls them, was not involved with the creation of the Christian Church

---

[a] This point cannot be emphasized enough through words and should be read with as much emphasis and annoyance as possible.

[b] The Cosmic Egg held everything within the universe before it was created. This story and the symbol are seen in Incan, Greek, Hindu, Dogon, Zoroastrian, Egyptian, Japanese, and Druidic creation stories.

[c] Bramley notes that the original Brotherhood of the Snake contained two serpents (the symbol of the caduceus) and focused on the scientific approach to spiritual understanding. After it was usurped by Enlil, there was only one snake (Bramley. *The Gods of Eden*, p. 55-66). This story and the separation between Brotherhood and White Brotherhood is also seen in *The Terra Papers*.

because it was interested in the works of all religions across all lands and strictly did not become part of any of them.[9]

Now that the Roman Catholic Church had its foundations and savior established, it was time to destroy the evidence and all other mystery teachings by any means necessary. Constantine denied state funding to pagan shrines, increased funding for the Christian Church, and created a tax-exempt system for clergy.[4] This caused a decline in paganism and a flourishing in Christianity.

To avenge the pagans, Constantine's nephew Julian ruled after Constantine died and attempted to reverse what had been done but was assassinated three years later while in battle against Persia. Christians within the Roman Empire became more ruthless in protecting their sacred Christian ideals. When Theodosius I ascended the throne, he issued eighteen laws to punish people who rejected the Christian doctrine. Heresy against the Christian church became treason as the Roman Empire descended into the Dark Ages.[4] The dawn of the ascending Kali Yuga was complete, and mankind was driven into complete materialism.

379-395 CE – Theodosius the Great, East Roman Emperor.

381 CE – Council of Constantinople, Second Ecumenical Council under Theodosius; Creation of the Trinity.

446 CE – The Brotherhood left Phoenicia in three different directions.[d] One group traveled to the Indus Valley, another to the Caucus Mountains, and the third arrived in Venice, Italy. In 1171, they became known as the Black Nobility after marrying into nobility and inventing or buying themselves titles.[10]

---

[d] David Icke notes that the Brotherhood are the rulers that came out of Babylon, work in secret, and are the power players behind every conspiracy theory (Icke. *The Biggest Secret*, p. 1).

. . .

527-565 CE – Justinian the Great, East Roman Emperor.

- Issued death penalty for heresy and pagans.
    - Ordered the burning of the Library of Alexander.
    - Hunting of heretics and pagans became a sport—burned heretics at the stake.

553 CE – Second Synod of Constantinople.

- Issued a decree banning past lifetime and reincarnation from Christian doctrine.[11]

570-632 CE – Mohammed and creation of Islam.

600s CE – The Khazar Empire was established in the northern region of the Black Sea in the Caucus Mountains—the possible origin of the Aryan race.[12] This empire was the second place the Black Nobility landed after they came out of Babylon.[11]

740 CE – Last battle between the Khazars and Muslims. The King, his court, and the military ruling class embraced Judaism and made it the state religion.[e,13] The lineage of Shem is associated with the Jewish lineage. The Khazars are possibly part of the lineage of Japheth.[14]

---

[e] The conversion of the non-Jewish lineage to being singly devoted to Jehovah is speculated by Koestler. He notes that one of the most significant motivations for this conversion was the Crusades and the constant battle between Christianity and Islam. Choosing Jehovah could have been a move to separate themselves from the battles (Koestler. *The*

. . .

748-776 CE[f] – Charlemagne introduced Germany to Rosicrucianism.[15]

THE CRUSADES AND PAPAL INQUISITIONS AND ORDERS

The Age of Pisces completed the decan of mutability and no longer needed to discover the type of Age it would become. As time moved into Pisces' fixed decan, the Crusades and Papal Inquisitions fulfilled the characterization of stagnation. War, death, and genocide in the name of the Piscean avatar who preached peace paired perfectly with the Iron Age mentality. While the Crusades were fought to save the Holy Land from Islam, it was a cover to serve a political purpose. Pope Urban II called for assistance from the Kings of England and France, as well as the German Emperor, to unite Europe under Christianity.[16]

900s CE – The Habsburgs (Hapsburgs) arrived in the region of Austria and governed the Holy Roman Empire until it was dissolved in 1806.

- Hapsburgs ruled Austria from 1287.
- In the 1500s, they inherited the crown of Spain.[17]
- The Habsburgs are stated to be descendants of Constantine.[18]

Thirteenth Tribe, p. 58-82). The interesting conspiracy of people who are Jewish by belief but are not of Jewish lineage pushes the antisemitism agenda as outlined by John Spargo in *The Jew and American Ideals*. This added to Koestler's knowledge of the Aryan race, and choosing Judaism shows a larger plan.

f Approximate date based upon established Rosicrucian active/inactive 216-year cycle. Lodge was established in Worms 1100 (active cycle), and the last active cycle would have been between 884 and 992. The previous active cycle was from 668-776. Charlemagne was King of Franks from 748-814. Leaving only 748-776 of the active cycle available.

1090 CE – Formation of the Assassin Order by Hassan Sabah within the Ismaili sect.

- Created the Judeo-Shi'a sect of Ismaili initiates.
- Followers divided:
  - "Self-sacrificers" were selected by the Lord, then made drunk and carried to a valley where he was "shown heaven and the seventy-two virgins" before he self-sacrificed.
  - "Aspirants" were allowed to move through the Seven Degrees of the Order.
- Used by Christian Princes to do their dirty work:
  - Richard of England and the death of Conrad of Montferrat.
  - Frederick II of Bavaria and the attempted assassination of the Duke of Austria. Innocent II excommunicated him for the attempt.[19]
- Corruption within the Order caused some to branch out and establish the Knights Templar.[20]

1091 CE – Count Elimar I founded the House of Oldenburg in Germany.

- From the House of Oldenburg came the kings of Denmark, Norway, and Hannover (Hanover, which was changed to Windsor in 1917).[11]

1099 CE[g] – Founding of Prieuré de Sion (Priory of Sion) by Godfrey of Bouillon, which was the Secret Order behind the Knights Templar

---

[g] This society has a website www.theprioryofsion.org and all the information contained on their website is consistent with information found in *The Holy Blood and the Holy Grail* by Henry Lincoln, Michael Baigent, and Richard Leigh.

and functioned as the administrative and military arm.[21] Existed to protect the Merovingian bloodline of Mary Magdalen, the bloodline of Jesus.[22]

- A secret Rosicrucian manifesto was written by German theologian and esoteric Johann Valentin Andrea, Grand Master of Prieuré de Sion, fiercely attacking the Catholic Church, Holy Roman Empire, and the Jesuits. The manifesto promised a world transformation through heretic principles and spoke of a time when the shackles of humanity would fall, humanity would liberate themselves and be the governors of their destiny, and it was all a part of the universal cosmic law.[23]

1100 CE – Rosicrucian Lodge was established in Worms, Germany.[15]

- 1100 – Nassau Castle was founded in the region of the Netherlands today.
- 1120 – Black Nobility Rupert I became Count of Nassau.
    - From Count Rupert I came the Hesse, Orange-Nassau, and Battenborgs Dynastys.[11]

1100s CE – The Black Nobility arrived in Florence, Italy. Notable families from this group are the de Medici, del Banco, and Agnelli.[11]

1066 CE – The Brotherhood from the Khazar Empire created the St. Clair/Sinclair lineage in Scotland.[24]

1118 CE – Knights Hospitaler was established after its initial formation as a charity group in 1048.

- After the Crusades, the Knights Hospitaler was disbanded. They became the Knights of St. John after the fall of Jerusalem in 1291. They were forced to leave and relocate. Those who moved to Rhodes became the Knights of Rhodes. Others moved to Malta, became the Knights of Malta, and became a key naval and military power in the Mediterranean, but were defeated by Napoleon in 1789.
    - Marino notes that the Order of St. John was in Genoa, Florence, and was influential during the time of Christopher Columbus.[25]
- In 1834, under Pope Leo XIII, they relocated to Rome and became known as the Sovereign and Military Order of Malta (SMOM). They are the world's smallest nation.
- They assist in anti-communist causes but are influential in politics, business, and intelligence within the US.[26]

1118 CE – Formation of the Knights Templar.

- Founded by Hugues de Payens with several initiates of the Assassin Order who followed Gnosticism in Palestine.
    - Degenerated after some members practiced the Phallicism of 'the Baphomet' (the idol of Luciferians). Miller notes that one of the rites of Templar initiation was the crime of Sodomy.
        - Heckethorn notes that there were controversies surrounding this accusation and that the Templars attempted to clear their name of this heresy. It was common at this time to bring incredulous charges against opponents.[27]
    - Thirteen Degrees – Degree 12 is called Knights Templar, and Decree 13 is called Knights of Malta.
    - Dissolved in England under Edward II.[28]
    - The Johannites were a branch of the original Knights Templar that strictly separated themselves from the

    charges. They followed the teachings of the Gnostics and became followers of St. John, also known as the Order of St. John.[25]

- 1317 – Became the Knights of Christ after King Dinis II of Portugal and Pope Clement V reformed the Order. Their emblem showed the Maltese Cross, commonly seen in reliefs of kings who were part of the Rosicrucian Order during the time of Ashurbanipal.
    - 1420 – King John I gave the Knights of Christ control of the Portuguese Indies.[29]

1146 CE – The Savoys of the Black Nobility ruled Italy until 1945. The Estes ruled Ferrara from the 1100s to 1860.[11]

1198 CE – Formation of the Teutonic Knights

- Teutonic Knights were formed strictly as a German Knight Order.
    - **1226** – German Emperor Frederick II restructured the order and made them overlords of Prussia.
    - **1234** – Knights only served under the authority of the Pope.
    - **Early 1300s** – Knight's domain extended to the southern coastline of the Baltic Sea.
    - **Early 1500s** – Knights "driven out" of Prussia by the Hohenzollern Dynasty of Poland.
    - The Austrian royal Hapsburg family sponsored the Teutonic Knights and became a secret Order under a different name.[30]

1179 CE – Alexander III called for a crusade against enemies of the Church and promised a reward of eternal salvation for those who died and two years of freedom from sins for those who fought.[17]

1204 CE – Pope Innocent III called for an attack on Constantinople by raping, pillaging, and burning the city. The attack was justified as punishment of the people by God for not submitting to the Roman Catholic Church.[17]

1208 CE – Pope Innocent III added to Alexander's rewards and offered the land and property of all heretics. He began the Crusade against the Cathars and Albigense.[17]

1229 CE – King Louis of France organized his first Inquisition to eliminate Cathars in France.

- **1247** – Massacre of Cathars in Montségur
    - Albigensian Crusade of Pope Innocent III: killed approximately one million people, including the Cathars and most of the population of South France.[17]

1329 CE – Merovingian line of France merged with the Stuart line of Scotland.[31]

1370-80 CE – The "Friends of God" movement started as a branch of the Waldenses in Switzerland, founded by banker Rulman Merswin. This movement stressed complete obedience to God and required followers to leave their past lives behind. The core teaching was that obeying God was the highest spiritual goal. Merswin received a revelation from a mysterious friend who helped him establish the movement and gave control to the Order of St. John. Bramley points out that this happened during a period when the Rosicrucians were active.[32]

1453 CE – Fall of Constantinople.

1469 CE – Ferdinand of Aragon and Isabella of Castile married.

1492 CE – Toledo, Spain became the new seat of Spanish Catholicism after the Moors were defeated in the fall of Grenada.

Unknown time during the Middle Ages – The Bauers family line of the Khazar Empire arrived in Austria and Bavaria.[11]

The religions of Islam, Judaism, and Christianity were firmly established, and Phase One of the Brotherhood's plan was complete. The conquest of Constantinople led to the Humanist Movement and sparked a new type of Renaissance as the land was no longer the sole source of capital for the Catholic Church. In the coming century, the bank would secretly take this land and show that it was more important than the country it "helped." The ending of the Crusades and Middle Ages marked another prophesied event and a fictional story—equal only to that of Moses and Akhenaten—was in the works. The Brotherhood could now move towards Phase Two: control the masses through religion.

Lies and the Age of Discoveries

The Age of Pisces had finally completed the fixed decan. At the turn of the 16th century, the Brotherhood had just started their ultimate plan of domination and control. They were aware of the greater mysteries of the universe but did not care, for they were solely focused on power—the exact thing that would ultimately become their downfall. Europe was under the dominion of the Brotherhood, but pieces of their plan were not set in stone. Yet.

The Creation of the Great Hoax of Columbus

The Story

The self-educated explorer Christopher Columbus left Spain in August 1492 in the name of the Holy Trinity with three ships: the *Nińa*, *Pinta*, and *Santa Maria*. He sailed towards the East Indies, hoping to profit from the Indian Spice Trade. He sailed under the patronage of King Ferdinand II and Queen Isabella I. Three months later, he landed in the Americas on the Bahamas coast.

What Actually Happened

The Set Up

Towards the end of the 15th century, prophecies of returning to Jerusalem (St. John's Apocalypse) were running high after the Inquisitions, the fall of Constantinople, the war of Grenada, and the Turkish territorial expansion. The predicted year for this prophetic return was 1500, the year of a new Jubilee. This marked the perfect time for the new Rosicrucian sect called the Society of Unknown Philosophers to act.[1] The Rosicrucian Brotherhood was aware of the Americas during the time of Thutmose III,[2] and the New World master plan was developed in Alexandria before the time of Mecca. Hall notes that the 'great discoverers' were mere agents of rediscovery for the Brotherhood.[3]

The Characters

Prince Nikolas Ypsilantis of the island Chios, Greece, was educated in the quadrivium, languages, and navigation.[4] He also studied Plato and other classical philosophers.[5] He was part of the Knights Templar, Society of Unknown Philosophers, the Franciscan Order Minims,[a] and a Rosicrucian, where the ties to the Rosicrucians are found in his cipher signature (Figure 35). He left 97 manuscripts and over 25,000 margin notes from his travels.[6]

In the stories of Columbus, John Cabot seems to disappear when Columbus appeared and vice versa. John Cabot was, in fact, Giovanni Caboto of Genoa and a citizen of Venice. Hall notes that Giovanni may have contacted secret sects like the Johannites. He partnered with the king of England, and Caboto's voyages allowed England and Spain to divide the Americas.[7]

---

[a] Founded in 1435 by Francis of Paola, it was a Roman Catholic Religious Order of Friars. It was founded in Italy but spread to Germany, France, and Spain.

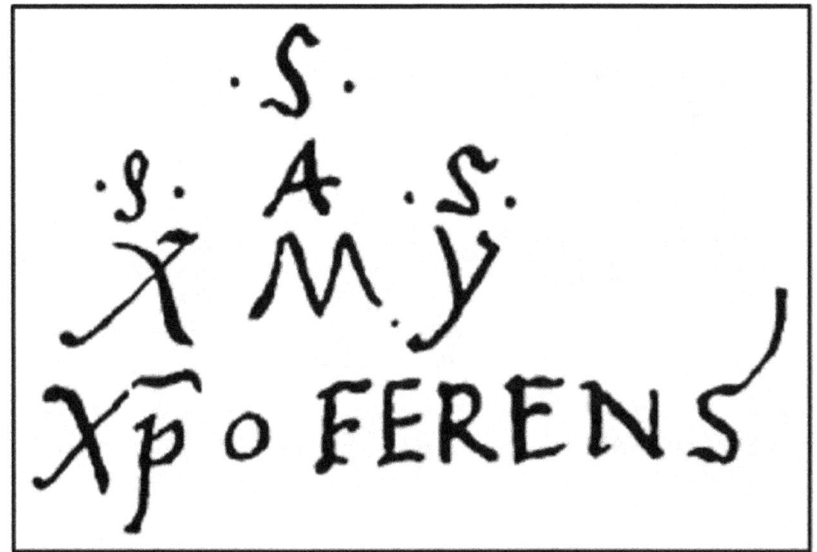

Figure 36: Christopher Columbus Signature

Credit: Public Domain,

The Financials

After Cosimo de Medici became the uncrowned monarch of Florence in 1434, the Medici family became head of the Italian banking House, led by Lorenzo de Medici, grandson to Cosimo.[8] Lorenzo did not live to see the Great Plan fulfilled. Through his belief and association with Secret Societies, he financially backed the fictional Columbus (actually Ypsilantis acting as Columbus) and helped him acquire more financial backing from other European institutions.[9]

. . .

The Timing

While the departure to the new world marked the promise of Jubilee and the prophecies, the arrival in the Yucatan region was equally as significant. The Yucatec Maya's hero-god was named Itzamna (or Zamna), the son of the holy one and supreme deity Hunab-Ku. When Itzamna originally arrived, he came from the east as a priest who helped heal the sick and restored life to the dead. [10]

Chapter twenty noted that Quetzalcoatl and Kukulkan were the same person, and Itzamna was another name for Thoth to those in the Yucatan region. Depictions show Itzamna rising from a serpent's mouth or turtle, symbolizing that he came from the sea. Before Itzamna died, he said he would return as a white-faced man.[10] The arrival of Columbus in the Yucatan aligned with the return of Itzamna, just as the arrival of Cortes aligned with the return of Quetzalcoatl.[11]

The Purpose

Just as the mysteries of Quetzalcoatl were known to the Europeans, they were also aware of the amount of silver and gold available in the Americas.[12] The Spice Trade was the cover used to hide the fact that the entire purpose was the wealth from the gold and the acquisition of the Americas to establish a new Philosophical Empire.[13]

The Spice Route

Columbus did not attempt to navigate to India for the Spice Trade. The Spice Route was a trade route that led from Europe to India, and during Columbus's voyage, it was impassable when he left Spain. The fall of Constantinople to the Turkish Empire caused the collapse of the trade routes, and the only Christian outposts left were Rhodes, Chios, and a small number of Greek Islands.[14]

The Name

Merriam-Webster notes that German cartographer Martin Waldseemüller named America. He wrote *America* on one of his maps in 1507 and named it after the Italian explorer Amerigo Vespucci. Unsurprisingly, this is false. The name comes from the Peruvian name for Quetzalcoatl, Amaru. *America* is derived from *Amaruca*, which translates as "Land of the Plumed Serpent," and the priests of this god (Amaru) ruled over both Americas.[15] The Merriam-Webster explanation is the standard story that is told and is a complete perversion of the truth behind what America was before the land was taken over, the gold was plundered, and millions slaughtered, all in the name of progress and achieving the dream of a few.

The Feathered Serpent's Appearance

Whether it was Quetzalcoatl, Kukulkan, or Itzamna, chapter twenty notes that this "mysterious figure" was Thoth. He was described as a white man with a broad forehead, large eyes, and a beard.[16] Recreating this type of figure played very well into the rediscoverer's hands.

Central and South America

1519 CE – Hernando Cortez lands in the area of Veracruz, Mexico, with approximately 600 men, and the Aztecs believe he is the reincarnated Quetzalcoatl. Cortez followed the same plan as Columbus and understood that when Quetzalcoatl left the Toltecs in Tula, he promised to return by sea and to the east. Ten cycles later,[b] Cortez arrived on the shores, and when Montezuma claimed their god had arrived, he did not deny it; instead, he leaned into it. When the Aztecs welcomed Cortez and his men into the palace, the Spanish seized Montezuma and held him for ransom. They demanded that the people bring enough gold to fill a ship.

Cortez planned to release Montezuma and place him on the throne as a

[b] The cycle taught to the Toltecs and Olmecs by Quetzalcoatl was 52 years (Sertima. *They Came Before Columbus*, p. 81).

puppet for the Spanish Empire until his second-in-command became restless and ordered the massacre of the commanders and noblemen. Their hunger for gold was quenched with 600,000 pesos of gold nuggets at the cost of bloodshed and slaughter.[17]

1530 CE – Francisco Pizarro landed in Peru with two hundred men. He planned to take over the Incan society the same way Cortez subdued the Aztecs. When his crew arrived, the Incas were in the middle of a civil war over succession. The challenger, Atahualpa, allowed Pizarro and his crew safe passage inland while he seized Cuzco. Pizarro then attacked their escort and held the challenger's son captive for ransom.

Church representatives and lieutenants placed Atahualpa on a mock trial and charged him with idolatry and the murder of his half-brother. From the arrival of Pizarro to 1750, over 100 million ounces of gold were taken from the Incan Empire.[18]

THE REFORMATION, PROTESTANTISM, AND THE JESUITS

1511 CE – Beginning of Rosicrucian active cycle of 108 years.[19]

1517 CE – Martin Luther posted his 95 theses and was excommunicated from the Roman Catholic Church.

- Luther was part of an Augustine Order[c] and developed ideas from the writings of Johann Tauler, who was associated with the Friends of God movement.

[c] Bramley (201) notes that Luther was a member of an Augustine Order that taught him the ideas from the Friends of God movement. Heckethorn (172) notes that the Templars followed the teachings of St. Augustine. There is an Augustinian Order founded in 1244 in Tuscany, Italy. The name changed to the Grand Union of the Order in 1256 and still exists in St. Thomas of Villanova (Augustinian.org).

- Disgusted by the Catholic Church and their relationship to the Medici family.
    - Pope Leo X was the son of Lorenzo de Medici. This mixing of money and the church led to spirituality becoming a business enterprise where Catholics could pay money to the church as compensation for sin.
- Taught that salvation was completely dependent upon the grace of God and the death of Jesus serves as penance of sin, completely wiping away the teaching of karma.
- Luther was supported by Philip the Magnanimous (Philip I), head of the Royal House of Hesse.[20]
- While there is no mention of his association with the Rosicrucians, Martin Luther's personal seal was a rose and cross.[21] He was also supported by members of the Rosicrucian Illuminati.[20]

1532 CE – Niccolò Machiavelli published *The Prince*, a novel from the Brotherhood that outlined their method for takeover and control. The principals included:

- Separate people into many competing groups.
- Create issues that cause fighting among the groups.
- Remain hidden as the instigator.
- Lend support to both sides of the fight.
- Allow the groups to view the secret instigators as benevolent problem solvers.
- Success is based on groups never knowing the true source of the problems.[22]

1533 CE – The Portuguese signed a trade contract with Grand-Popo West Africa.[23]

- This treaty began trade with Africa, leading to the Atlantic slave trade and Dahomey Kingdom.

1536-1538 CE – John Calvin and the Puritan Movement Teachings:

- Salvation was pre-determined by God before birth, and there was nothing that anyone could do if they were not "chosen."
- The elect few who were chosen served God by suppressing those who were not saved.
- Calvin also taught that mankind is still being punished for Original Sin.
- Tension and struggle were constant in life because of the eternal battle between God and Satan.
    - This teaching led to the belief that peace means that Satan has won.
    - Going to war was to fight for the glory of God.
- His teachings created philosophical feudalism.[24]
- David Icke notes that John Calvin was a fake name. His real name was Jean Cauin, from Nyons, France, and he attended Collège de Montaigu,[d] the same as Ignatius Loyola. After Cauin had completed his education, he moved to Geneva, Switzerland, and changed his name to Cohen. It was in Geneva that the philosophy of Calvinism was developed. He changed his name to John Calvin before moving to England and preaching a literal version of the Ten Commandments.[25]
    - The doctrine of Calvinism was part of the Black Nobility's secret movement towards a banking empire. With this plan came the planning of the entire 16th century.[23]

1540 CE – Order of the Jesuits established by Ignatius Loyola. [26]

---

[d] Icke notes that the Collège de Montaigu was controlled by the Brotherhood.

- Consistently in opposition to the Church and focuses on power and wealth over the things of Christ.[27]
- The most important thing within the Order was unquestionable obedience free of reason.[28]
- The Order of the Jesuits contained Thirteen Degrees, with the highest called the General, also known as the Black Pope.[24]
    - The Pope was subservient to the General.[24]
    - The Jesuits have been consistently banned from every country they had power.[29]
    - The University of Ingolstadt was the Jesuit University where Adam Weishaupt was a teacher and began the Bavarian Illuminati.[30]

Sir Francis Bacon 1562-1626

It is impossible to explain the changes from 1500 to 1700 without mentioning Sir Francis Bacon. His life story reads like someone who lived at least two lives. He was the illegitimate son of Queen Elizabeth and Robert Dudley, Earl of Leicester, who rose to become Lord Chancellor of England and Grand Master of the Rosicrucians of England. He was the real author of Shakespearian works and hid within the plays secret Rosicrucian teachings and the actual Freemasonry rituals. He remodeled modern law and was the editor of the Bible under King James I of England. Manly P. Hall notes that he spent a year editing the Bible to encrypt secret Rosicrucian keys of mystic and masonic Christianity into the teachings.[31]

He wrote *The Fama Fraternitatis* and *New Atlantis*. He worked secretly through the underground channels of the Inns of Court— the center of Brotherhood legal professions—during his apprenticeship with Philip II of Spain to win North America for England.[32] He worked with the Society of Unknown Philosophers to create a philosophical empire of Freemasonry in America.[33]

He was the father of modern science in the Scientific Revolution and Age of Enlightenment and taught that the world is all there is.[34] He worked with others within the Society of Unknown Philosophers from

Germany, France, and the Netherlands to create a plan of indoctrination within the Colonies to create the atmosphere for a revolution 100 years before the Revolutionary War.[35] He was one of the founders of Freemasonry in America but did not create Freemasonry; he linked it to a much older tradition,[36] possibly the Dionysian Artificers of Hiram Abiff.

While Sir Francis Bacon's achievements were substantial, it is not the person that is of real interest. His life was a focal point between the invisible Rosicrucian Brotherhood and the true esoteric magicians who create societal mutations. He did all this while flawlessly executing his place in the Black Nobility Stuart lineage from Elizabeth I. He would have made Niccolo Machiavelli proud as he was the Grand Master of the Rosicrucians and worked to divide the church while editing the King James Bible in 1611. He practiced mystical esotericism while preaching scientific materialism.

1569 CE – Eighty Years War began between the Calvinists in the Habsburg Netherlands against Spain.

THE SEVENTEENTH CENTURY AND THE AGE OF ENLIGHTENMENT

May 14, 1607 – Arrival at Jamestown

1609 CE – Truce made between Spanish and Habsburg Dutch to suspend fighting for 12 years. Truce marked Dutch independence, and the Bank of Amsterdam was founded.[37]

1619 CE – The Bank of Hamburg established.[37]

1642-1668 CE – The German takeover of England.

- After a series of takeovers, wars, a plague, and the Great Fire of London, William III, Prince of Orange, replaced the Stuart Dynasty. He was only the interim ruler until the Hanoverian king George Louis became King George I of England.[38]
- The Great Fire of London allowed the Brotherhood to unite in London. They used the freshly cleared land from the fire to build the Bank of London in 1694. It marked the shift when the bank became more important than the country.[39]

1661 CE – The Bank of Sweden established. [37]

1694 CE – The First Rosicrucian colony was established by Johannes Kelpius in Philadelphia.[40]

The 17th century ended as the long-planned actions of the Brotherhood began to show. Europe's banking system united in the Brotherhood stronghold of Geneva, Switzerland. Their threefold plan of controlling the politics, religion, and financial institutions was well underway. As the new century approached, the Brotherhood looked toward the West for expansion of their secret empire.

The New, Shiny England and Move Towards Domination

After completing the Kali Yuga of endless war, time moved into the dawn of the Dvarapa Yuga as the Piscean cardinal decan showed the immutability of universal laws. The Brotherhood would not give up their dominance and became stronger than ever in the 18th century. The aftermath of the Reformation, Puritanism, and the Eighty Years War created the ideal world for the Brotherhood. As the pawns moved across the chessboard, the Brotherhood quietly planned and enacted their checkmate strategy.

Eighteenth Century

Europe

1716 CE – The Bank of France established.

European Freemasonry

Freemasonry was initially separate from political or social issues and purely for trade guilds. In 1700, Freemasonry underwent a complete overhaul in England three years after the Hanoverian King George Louis was coronated and ruled as King George I of England. The Mother

Grand Lodge of the World was established and supported the rule of the German Hanoverian line.[1]

Freemasonry became two distinct sects as the common European Freemasonry became known as Scottish Freemasonry (Scottish Rite) and the Grand Orient. In Italy, the Grand Orient network of Freemasonry was formed in 1733 at the Black Nobility stronghold in Florence. By the time the Revolutionary War began, every country was part of a Freemasonry society within the Brotherhood.[2]

After the Hanoverians took over the English crown, they did not trust the people they ruled for fear of the Pro-Stuart subjects taking over. George I used money from the British treasury to rent mercenaries from his lineages' region of Hanover in Germany.[3] This increased the wealth of the House of Hanover and Hesse during each English war.

1764 CE – Formation of the Order Strict Observance.

- Elite Freemason group that included the Hesse Dynasty.[4]
- Recruited members from Lodges and focused on formulating a grand plan for economic rule.[5]

1760s CE – Martinism founded.

- Founded by a Marrano Jew named Martinez Depasquale, who preached the cult was the advancement of man's spiritual nature and reintegration of the Brotherhood inherited from Eli, Enoch, and Melchizedek.[6]
- The real purpose of Martinism was to unite people for political schemes to overthrow the monarchy.[5]

1765 CE – Mayer Amschel Bauer of the Khazar Dynasty becomes Mayer Amschel Rothschild in Frankfurt and partners with William IX of the House of Hesse.[7]

- 1789 – Rothschild takes over as William IX's financial agent.[8]

The Bavarian Illuminati

It is a little-known fact that the Illuminati is part of the Rosicrucian Order. Pythagoras was initiated into the Illuminati in 529 BC in Thebes.[9] Many secret societies have degrees or rites named Illuminati or Illuminated Ones, but the one that would change the world known today was founded in 1776 by Adam Weishaupt. While formulating his master plan, he attended Freemasonry meetings to learn about the Order and ultimately joined the Strict Observance in 1777. Weishaupt and the five original members infiltrated Freemasonry and Strict Observance to recruit for the Illuminati. Within the first three years, it spread to four cities in Bavaria and included 54 noblemen and clergy.[10]

As the first society to strictly exist for political means, it spread quickly throughout Germany and Austria, with 300 members by 1782. The Illuminati were Martinists and spread throughout various Orders and became hidden within all secret societies. The objectives of the Bavarian Illuminati included:

1. The destruction of Christianity and all monarchies.
2. The destruction of sovereign nations.
3. The discouragement of patriotism and national loyalty.
4. Elimination of family ties.
5. Overthrow of personal rites to inheritance and property.

The various Orders and players worked together to create destruction via financial, intellectual, and anti-Christian means, ultimately leading to the creation of the French Revolution.[11]

Freemasonry and the Founding Fathers

After the founding of the initial colonies, the tension between France and England was high over colonization in the New World, and two wars were fought. Uneasy during peaceful times, England sent George Washington—initiated as Freemason in 1752 and a loyal British military officer—to keep the troops of France from taking land belonging to England. The French and Indian War began and became the Seven Years War. Taxes increased, and the build-up is history. What is not common knowledge is that every step leading the Colonists into revolution was strategically made by the Brotherhood through the various arms of Freemasonry already established by Sir Francis Bacon.[12]

The instigator of the Revolutionary War is said to be King George III. History shows that he was friends with the Colonists. The real driver behind the future revolution was found in the Freemasonry Lodges of the Colonies.[13] Benjamin Franklin was already a Masonic Grand Master and the Order of the Quest's appointed spokesman from France. He subtly influenced the Colonists with the writings of *Poor Richard's Almanac.* Franklin was honored by the most famous French secret Order, Lodge of Perfection, for his part in the American Revolutionary War and the founding of America.[14] Other Masonic Grand Masters included Washington, Paul Revere, John Hancock, and James Clifton. Masonry became so prevalent that they made one-seventh of the Continental Army.[15]

France and Prussia supported the Colonists and sent militia. However, the overarching Hanoverian contract between England and Prussia meant 30,000 German soldiers fought for England. William IX of Hesse-Cassel provided 15,000 soldiers, and he doubled his wealth from the war.

After America won its independence, Alexander Hamilton became the director of the Bank of New York in 1784 and repeated the model set by the Bank of England in 1694. He convinced the young government of America to turn all state debts into one national debt, place it in a privately owned central bank, and only pay the interest. The Central Bank of the United States was established in 1791. The United States does not look exactly like Great Britain today because of George Mason, the divine spark of humanitarianism from the Brotherhood. He detested every move by the new government and pushed for the Bill of Rights.[16]

CHRISTIANITY AND AMERICAN SLAVERY

Missionaries and the Creation of Racism

The main mission of white Americans when missionaries came to spread the word of Jesus to their slaves was to make the slaves more content in their enslavement. Missionaries and Preachers handpicked scriptures about obeying their master and the promise of heaven after service and good works on Earth. This led to African Christianity and the language of the Black Church's spirit of survival.[17]

The problem with slavery and Christianity co-existing was that the Euro-Americans had to justify and define their meaning of 'race,' 'human,' and 'morality.' To make the active disassociation easier, the main idea became that 'European' meant 'Christian,' which looked white to Europeans, while 'African' meant 'heathen.' There was even a question over whether or not Africans had a soul.[18]

The Brotherhood's stench was all over the Slave Trade and slavery in America because it was specifically designed as a racial form of bondage. Christianity was central to creating and imposing racial categories in society. In the 18th and 19th centuries, racism and religion defined civilization by whitewashing African and Native American culture with Christianity for the sake of culture. Racism entered the 'New Atlantic Utopia' that Bacon had dreamed about, bringing the worst out of people.

Africa

The Slave trade in West Africa was growing throughout the 16th and 17th centuries. The Dahomey tribe had gained dominance within the region under the rule of Agaja. In 1727, the Dahomey Kingdom overthrew the coastal kingdom of Hueda and seized possession of the port city. This action caused the kingdom to flourish at the expense of the country.[19]

The Nineteenth Century

The French Revolution caused by the Bavarian Illuminati caused republican-style governments in Europe, Africa, Russia, and South America. After war and successive revolts lasting from 1795 to 1830, the banks took a turn for the worse and there was an international credit collapse in 1848.[20] Once again, the tides of empires changed as Hesse-Cassel and Rothschilds grew in wealth and influence, along with the Black Nobility, and the people continued to suffer.

Karl Marx emerged and promoted the ideals instilled in him through his German occultist teacher, Bruno Bauer, a relative of Rothschild.[21] The teachings of Marxism show that the Brotherhood was attempting everything it could to hold onto their Piscean age as long as possible and continue in the materialist Kali Yuga. With Marxism in 1833 came the development of psychology in 1879. Both focused on scientific reasoning, materialism, and denied the existence of the supernatural.[22]

The Order of the Golden Dawn was created in 1888 by W. Wynn Westcott, a member of the Rosicrucians.[23] This Order was to be the Masonic-Rosicrucian sect only for Masons truly interested in learning the ancient teachings.[24] It was a blend of ancient Druidism, black magic, and blood ritual. There were rituals to access full psychic potential and contained information about Vril and the secret underground Brotherhood, with the society's secret sign: the salute of 'Heil-Hitler.'[23]

1863 CE – The Federal Reserve Bank was established.

. . .

1865 CE – Formation of the Ku-Klux Klan in Tennessee.

- Grew through 1865-1868 when it absorbed aligned societies called Knights of the White Camelia, the White Brotherhood, Black Calvary, and White Rose.
- Principles of maintaining peace and order with the supremacy of the white race and intermingling races.
- It was forced to break apart in 1871 and 1872, and the central hub of the Klan dispersed. But in 1915, the Klan was revived.

The last part of the Brotherhood's plan before their checkmate was to trial their greatest weapon, the concentration camp. Cecil Rhodes was an Englishman who lived in South Africa and exploited their mineral resources. He wanted to build a political system with a universal government led by Britain. This desire caused him to create the Round Table, backed by Lord Rothschild and Alfred Milner. During the Second Boer War in 1900, concentration camps were created for Dutch settlers and killed over 20,000 people. In Great Britain, the Round Table became known as the Royal Institute of International Affairs, and in the US, it is known as the Council on Foreign Relations.[25] Everything was in place as the Brotherhood's Messiah sat in a bookstore learning about German mysticism.

Project Domination

The Brotherhood had successfully used all the tactics of Machiavelli while remaining the shadowy puppet masters behind it all. Phase Three was running as planned, and no one was the wiser. Yet they had a secret in their metaphorical back pocket waiting to be dropped like a royal flush with the highest stakes.

After the assassinations of Franz Ferdinand and Dutchess Sophia of Austria, World War I began. World War II quickly followed as the Nazi party attempted their domination via the Third Reich. The Black Nobility and Khazar Empire houses backed both sides of the war to create their ultimate monopoly over every industry. Actions by the Nazi party during and after the war led to the question of whether or not the Aryan race and German dominance were the ultimate purpose of World War II.

1923-1933 CE – The Germans learn of a secret underground reptilian base in Antarctica.

1938 CE – The Nazi Party begins exploring settlement.

. . .

1939 CE – Ships leave from Germany for Antarctica with scientists and unusual equipment.

- The UK learns of Nazi base construction in Antarctica.
- 1939-1949 – Haunebu Craft I-IV are created.[1]

1940 CE – Nazi Party begins building Neuschwabenland (New Swabia) Aryan colony in Antarctica.

1941 CE – German-run company I.G. Farben infiltrates American companies and successfully establishes the Fourth Reich in America.

1942 CE – Roosevelt overthrew the U.S. front organization Union Banking Corporation in New York because it was a Nazi company with Prescott Bush as founding member and director.

1945 CE – Hitler and Eva Braun escape to Antarctica. Nazi's surrender and WWII is over.

- Haunebu III travels to Mars.
- Project Overcast begins recruiting Nazi scientists to bring them into the U.S. It is later dissolved and changed to Project Paperclip.

1946 CE – Operation Highjump leaves for Antarctica.

- 1947 – Operation arrives in Antarctica and lasts only two weeks instead of the planned six months.
- Truman creates:
    - Majestic-12
    - CIA
    - NSC
    - NSA

1947 CE – Roswell UFO crash.

1952 CE – Project Blue Book is renamed from Project Sign (1947) and Project Grudge (1949) begins UFO investigation and obtains 3200 cases by 1954.

- Began under Captain Edward J. Ruppelt
- 1958-1963 – Captain Robert J. Friend of the Tuskegee Airmen[2] placed over Project Blue Book. He attempted to return Project Blue Book to its original purpose with updated files, catalogs, and observational statistics. His story is suppressed.

1953 CE – Kingman, AZ UFO crash.

1955 CE – Eisenhower meets with the EBEn (Extra-Biological Entity).

1958 CE – NASA established.

1962 CE – Kennedy forms the Defense Intelligence Agency (DIA).

. . .

1961 CE – Betty and Barney Hill abduction.

- Betty's focus:
  - Betty told her sister Janet Miller over the phone about the event.
  - Janet spoke with a neighbor whose husband was a physicist and was informed by the former chief of police in the neighboring town that all UFO sightings should be reported to Pease Air Force Base.
  - Consistently wanted to tell people and get their story out.
  - Focused on finding answers and searching for the truth about what happened.[3]
- Barney's focus:
  - After the incident, Barney suggested they go into separate rooms and draw their observations.
  - He then told Betty they should refrain from telling anyone because they would never be believed.
  - When he spoke with Pease Air Force Base, he left out the observation of humanoid figures because he did not want to be seen as crazy.
  - He wanted to stay quiet and focus on things in his life that had meaning, such as the Civil Rights Movement and community involvement.[3]

1965 CE – Ships connected to EBEn from Serpo land and Project Crystal Knight begins.[4]

Key Nazi Players responsible for Project Paperclip and establishing the Fourth Reich in America:

- Allen Dulles

- John Foster Dulles
- Lt. Colonel Reinhard Gehlen[5]

Secret Projects that continue without the knowledge of the public:

- Jump room capabilities to the Mars base[6]
- Secret Moon base
- Space station that orbits Jupiter
- Secret human slave trade
- Space Nazis
- Twenty and Back program

There are so many unknowns that exist in today's world. Secrecy has been the name of the game for thousands of years and has cost humanity everything. As time moves towards the Age of Aquarius, the moment has come to remove the stigma and fear of ridicule from the Civil Rights Movement during the massive contact campaign by humanity's Extraterrestrial Ancestors. No one should be able to tell another what to do or who to be. Only you can decide to change the narrative in the last act of this grand play called Life.

THE BROTHERHOOD'S UNDOING

However distorted the mysteries became throughout the age, the original mysteries of the Vedas, Mazdeism, and Egyptian mysteries were the purest compass towards a life of reason and virtue, and most importantly, they were available to all. The separation between those deemed worthy and unworthy of being initiated into the esoteric mysteries is the greatest robbery of mankind by those who called themselves Aryans. While history has shown that no purity remains long in religion, the Brotherhood used their methods of domination and control flawlessly. They effectively destroyed culture and knowledge at every turn. Just like the HEN-T used the mind-control techniques of the ARIAN Queens, mankind lost the esoteric knowledge and all memories, understanding, and knowledge of their takeover. Christianity, Judaism, and Islam had won as the ultimate vehicle for erasing vital information that humanity desperately needed.

Through racism, religious distortion, and doomsday prophecies, the Brotherhood successfully controlled society. Mankind was kept in an endless cycle of global change and fighting for the fake problems created by the Brotherhood to continue their cycle of problem-solution-hero. The build-up from the time of Constantine to the unification of Europe, Russia, Africa, America, and South America under the banking empire led to the Brotherhood's next conquest: the complete domination of the human race repackaged under the guise of freedom. By the end of the cardinal decan in the Piscean Age, the Brotherhood was white-knuckling control as the veil of humanity's consciousness began to secretly lift.

Throughout history, the silent and invisible good guys waited for their moment to awaken the people. They used the advanced technology rediscovered during World War II and subsequent Operations and Projects against the Brotherhood. The blanket of suppression was ripping as humanity began speaking of the Brotherhood and other benevolent extraterrestrial beings they once called friends. As the tools of the Brotherhood began to crumble, the collective voice of humanity grew. Truth rang louder and the light of divinity brought out the darkness of control. The transition into Dvapara Yuga and the Age of Aquarius performed a type of mystical alchemy in humanity's consciousness equal to that of the last mystery teaching.

The people began speaking of every spiritual and extraterrestrial encounter. The voices of the most victimized groups were heard as the cracks of racism, stigma, and religion that once created unbreakable barriers crumbled. Together, mankind became their own savior and saved themselves from the grip of the Brotherhood evil named Jehovah.

The major voices of this movement were empowered and left their mark within the annals of history. The voices from the B.L.A.C.K. Community included...

...yours.

If there is a story you would like to share with Project B.L.A.C.K., please call 888-345-9121.

I look forward to hearing about your unique experience. All questionable experiences are welcome.

Afterword

The ancient traditions remind humanity of our origins. We're beginning to wake up, remember the ancient teachings, and come out of the cycle and lies that have held us down for so long. We're starting to remember that humanity's ancient ancestors came from the stars. The time is now to tell your story to help others tell theirs. The Black Community has been held down by the Christian Church for too long, and it has kept us blinded to things in front of our faces.

Too many voices have been missing from this grand story for far too long. It is time to reawaken to the stories of the Dogon, Olmecs, and Native Americans and show that they are here as ancient friends, and it is time to step up and create an atmosphere filled with the voices of humanity. Yes, the black community is different. Yes, we have been through difficulties and suppression. Yes, it is time to leave those things in the past and shout your story from the mountaintops that the Brotherhood tried to push you away from.

While the climate of Ufology is beginning to shift, the change is in the toddler stage and hesitant with new things. After learning the true history of humanity, I hope that while it is shocking, it helps you understand that the time has never been better than now.

Project B.L.A.C.K. is a movement to empower the black community to openly discuss and explore the important topics of extraterrestrial life and UFOs. The Mission of Project B.L.A.C.K. is to break down barriers and promote education, understanding, and empowerment about extraterrestrial life and UFOs in the black community by "Starting the Conversation" and breaking down the barriers of repression, religion, taboo, stigma, and skepticism.

Notes

1. Ancient Antiquity

1. *Chapter One*
   Thomas. *The Adam and Eve Story.*
2. Hapgood. *The Earth's Shifting Crust.*
3. "Moai Megalithis of Easter Island."
4. "Smithsonian Cover-Up: Ancient Egyptians and Giants in the Grand Canyon."
5. Thunderbird. *Revealing America's Dark Skinned Past Vol 1*, p. 13-15.
6. Sertima. *They Came Before Columbus*, p. 32-33
7. 00:25-04:50 "Layers of Machu Picchu"
8. 07:33- 08:12 "Megalithic Mysteries of Lake Titicaca"
9. "Göbekli Tepe and the Prophecy of Pillar 43Apocalypse and the Vulture Stone"
10. "Underground Cities of Cappadocia"
11. 01:43-03:07 "Denisovians: They Might be Giants"
12. Gaur, A.S, et al. "An Ancient Harbour at Dwarka"
13. 01:13- 03:30 "The Sunken Kingdom of Krishna"
14. "The Band of Peace"
15. Tellinger. *Temples of the African Gods*, p. 32
16. Ibid, p. 47-50
17. Ibid, p. 64
18. Ibid, p. 46
19. Ibid, p. 122-131

2. Forgotten Inhabitants

1. *Chapter Two*
   "Forbidden Archaeology: Lost Giants of AmericaThe Smithsonian's Biggest Secret."
2. 05:10-06:33 "Elongated Head DNA Discovery."
3. 12:08-13:02 Ibid.
4. Zhang, Qun, "Intentional Cranial Modification from the Houtaomuga Site in Jilin, China."
5. 13:30-14:16 "Elongated Head DNA Discovery."
6. 14:18-19:28 Ibid.
7. www.loughcrewmegalithiccentre.com/loughcrew-cairns/

3. Advanced Knowledge: Star Alignments of Megaliths

1. *Chapter Three*
   Maxwell. *That Old-Time Religion*, p. 4.
2. Bauval. *The Orion Mystery*, p. 116.
3. Temple. *Sirius Mystery*, p. 88.
4. Ibid, p. 63.
5. Herschel. *Hidden Records Vol 1*, p. 21.
6. Reedijk.
7. Agius, George.
8. Reedijk.
9. Herschel. *Hidden Records Vol 1*, p. 14.
10. Ibid, p. 25.
11. Tellinger. *Temples of the African Gods*, p. 30.
12. Herschel. *Hidden Records Vol 1*, p. 59.
13. David. *The Orion Zone*.
14. Biglino. *Gods of the Bible*, p. 126-127.
15. Norris. "The World's Oldest Story?"
16. Ibid, p. 63-64.
17. 03:22-04:06 "Deep Space."
18. 15:20- 17:46 "Layers of Machu Picchu."
19. 04:56-05:24 Ibid.
20. Biglino. *Gods of the Bible*, p. 126.

4. Ancestral Narratives

1. *Chapter Four*
   Scranton. *Sacred Symbols of the Dogon*, p. 30-38.
2. Temple. *The Sirius Mystery*, Ch 9.
3. Waters. *Book of the Hopi*, p. 3-27.
4. 04:08-04:39 "The Pleiadians."
5. Morningstar. *The Terra Papers.*
6. Mullins. *Star People, Sky Gods and Other Tales*, p. 11.
7. 06:17-06:51 "The Pleiadians."
8. Mullins. *Star People, Sky Gods and Other Tales*, p. 41-42.

5. The Bible and The Sages

1. *Chapter Five*
   Biglino. *Gods of the Bible*, p. 69.
2. Blavatsky. *The Secret Doctrine Vol I*, p. 244.
3. O'Brien. *The Genius of the Few*, p. 30-31.
4. Dalley. *Myths of Mesopotamia*, p. 292.
5. Schwemer. "Bīt mēseri at Aššur", p. 64-66.
6. Reiner. "Etiological Myth of the 'Seven Sages'."

7. Hess. "I Studied Inscriptions from Before the Flood", p. 232.
8. Smith. *Chaldean Account of Genesis*, p. 28.
9. LaCroix, Matthew. "Keepers of Knowledge."
10. Lambert. *Babylonia Creation Myths*, p. 209.
11. 21:13- 23:10 "The Layers of Machu Picchu."
12. O'Brien. *The Genius of the Few*, p. 27.
13. Morfill. *The Book of the Secrets of Enoch*, p. 2 v.5
14. Doreal. *The Emerald Tablets of Thoth-The-Atlantean*, p. 1.
Part One Summary

Part I Summary

1. Durvasula. "Recovering Signals of Ghost Archaic Introgression in African Populations."

6. Changing the Biblical Narrative

1. *Chapter Six*
Jacobsen. *Sumerian Kings List.*
2. Dalley. *Myths of Mesopotamia*, p. 9-35
3. Sitchin, *The 12th Planet,* p. 350-355
4. Ibid p. 357
5. Sitchin. *Genesis Revisited,* p. 170-171
6. Sitchin, *The Lost Book of Enki*, p. 197
7. *Bible.* Authorized King James Version
8. 43:32-43:37 & 56:09-57:30 "Darwin Evolution Theory Debunked by Billy Carson"
9. *Code of Hammurabi*, p. 26.
10. Clark. *Etymological Dictionary.*
11. Biglino, *Gods of the Bible*, p. 77-78
12. Sitchin. *Lost Book of Enki*, p. 248
13. Biglino, *Gods of the Bible*, p. 80.
14. Desmarquet. *Thiaoouba Prophecy*, p. 30-40
15. Smith. *Chaldean Account of Genesis*, p. 51 line 18.
16. Ibid, p. 53
Chapter Seven
17. Morningstar. *The Terra Papers*, p. 38-9.
18. Bramley. *The Gods of Eden*, p. 66.
19. Lewis. *Rosicrucian Questions and Answers*, p. 33-42.
20. Bramley. *The Gods of Eden*, p. 66.

7. The Beginning of the Biblical Patriarchs

1. Kramer. *Myths of Enki, The Crafty God*, p. 31-37
2. "Book of Enoch" *The Complete Apocrypha,* p. 196-237
3. Morfill. *The Book of the Secrets of Enoch*, p. 1-5.

4. Daniel 4:13, ESV
5. Biglino. *Gods of the Bible*, p. 213.
6. Sitchin. *The Wars of Gods and Men*, p. 38.
7. Sitchin. *The Lost Book of Enki*. p. 173-194
8. Ibid, p. 189
9. Biglino. *The Gods of the Bible*, p. 103.
10. Ibid, p. 186-189
11. Ibid, p. 190-191
12. Lambert. "Enmeduranki & Related Matters" *Babylonian Creation Myths,* p. 130
13. Sitchin. *Genesis Revisited*, p. 206-207.
14. Biglino. *Gods of the Bible*, p. 108
15. Kramer. "The Sumerian Deluge Myth" *Selected Writings of Samuel Noah Kramer*, p. 119-120.
16. "Book of Enoch", p.197
17. Sitchin. *The Lost Book of Enki*. p. 195-216
18. Ibid, p. 200-203
19. Ibid, p. 203-205

8. Before the Eighteenth Dynasty

1. *Chapter Eight*
   Josephus. *Against Apion I*, 1.73, 82-87.
2. Shaw. *The Oxford History of Ancient Egypt*, p. 160-61.
3. Ibid, p. 174.
4. Josephus. *Against Apion I*, 1.75-90.
5. Shaw. *The Oxford History of Ancient Egypt*, p. 180.
6. Ibid, p. 183.
7. Ibid, p. 201-202.
8. Ibid, p. 177.
9. Sitchin. *The Cosmic Code*, p. 95, 109.
10. Shaw. *The Oxford History of Ancient Egypt*, p. 177.
11. *Ugaritic Texts Ba'al Cycle*, p. 3.
12. Sitchin. *The Stairway to Heaven*, p. 198-219 and *Ugaritic Texts: Ba'al Cycle* [entire book].
13. Sitchin. *Stairway to Heaven*, p. 201-205.
14. Ibid, p. 206-210.
15. *Ugaritic Texts: Ba'al Cycle*, p. 9.
16. Ibid, p. 19-20, 42.
17. Sitchin. *Stairway to Heaven*, p. 214-219.

9. The Eighteenth Egyptian Dynasty

1. *Chapter Nine*
   Josephus. *The Complete Works,* Apion 1:9

The Eighteenth Dynasty

1. Sitchin. *The Wars of God and Men*, p. 39-40
2. Josephus, *The Complete Works*. Apion,1:81-92
3. Osman. *The Egyptian Origins of King David and the Temple of Solomon*, p. 35.
4. Shaw. *The Oxford History of Ancient Egypt*, p. 207.
5. Ibid, p. 208.
6. Shaw. *The Oxford History of Ancient Egypt*, p. 209
7. Lewis. *Rosicrucian Questions and Answers*, p. 43-44.
8. Shaw. *The Oxford History of Ancient Egypt*, p. 214.
9. Lewis. *Rosicrucian Questions and Answers*, p. 44.
10. Shaw. *The Oxford History of Ancient Egypt*, p. 220
11. Osman. *The Egyptian Origins of King David and the Temple of Solomon*, p. 36
12. Breasted. *Ancient Records of Egypt Volume II: The Eighteenth Dynasty*, 98
13. Shaw. *The Oxford History of Ancient Egypt*, p. 221-222
14. Breasted. *Ancient Records of Egypt Volume II: The Eighteenth Dynasty*, 91-98
15. Ibid, 115-122.
16. Gardiner. *The Egyptians*, p. 178
17. Shaw. *The Oxford History of Ancient Egypt*, p. 228
18. Osman, *The Egyptian Origins of King David and the Temple of Solomon*, p. 37
19. Gardiner. *The Egyptians*, p. 178
20. Shaw. *The Oxford History of Ancient Egypt*, p. 241.
21. Osman. *The Egyptian Origins of King David and the Temple of Solomon*, p. 37-38
22. Lewis. *Rosicrucian Questions and Answers*, p. 44-53.
23. Breasted. *Ancient Records of Egypt Vol 2*, 780
24. Shaw. *The Oxford History of Ancient Egypt*, p. 241, 245-6
25. Ibid, p. 241-2
26. Breasted. *Ancient Records of Egypt Vol 2*, 804
27. Ibid, 815
28. Shaw. *The Oxford History of Ancient Egypt*, p. 248-9
29. Breasted. *Ancient Records of Egypt Vol 2*, 626-28, 830-38
30. Ibid, 833
31. Shaw. *The Oxford History of Ancient Egypt*, p. 250-51
32. Björkman. "Neby, the Mayor of Tjaru in the Reign of Thutmose IV", p. 51
33. Osman. *Moses and Akhenaten*, p. 121

10. The Amarna Kings

1. *Chapter Ten*
   Gardiner. *The Egyptians*, p. 200
2. Shaw. *The Oxford History of Ancient Egypt*, p. 253-259
3. Ibid, p. 265-267
4. Breasted. *Ancient Records of Egypt: Vol II*, 869.
5. Lewis. *Rosicrucian Questions and Answers*, p. 54-55.
6. Osman. *Moses and Akhenaten*, p. 106-116
7. Dodson. "Crown Prince Djhutmose and the Royal Sons of the Eighteenth Dynasty," p. 88.

8. Osman, *Moses and Akhenaten*, p. 122.
9. Lewis. *Rosicrucian Questions and Answers*, p. 55-56.
10. Shaw. *The Oxford Ancient History of Egypt*, p. 267-68
11. Lewis. *Rosicrucian Questions and Answers*, p. 57.
12. Ibid, p. 269.
13. Osman. *Moses and Akhenaten*, p. 123.
14. Bramley. *The Gods of Eden*, p. 69-70.
15. Lewis. *Rosicrucian Questions and Answers*, p. 60.
16. Osman. *Moses and Akhenaten*, p. 62.
17. Ibid, p. 127.
18. Shaw. *The Oxford Ancient History of Egypt*, p. 270
19. Bennett. "The Restoration Inscription of Tutankhamun", p. 9.
20. Lewis. *Rosicrucian Questions and Answers*, p. 58-59.
21. Bramley. *The Gods of Eden*, p. 68-69.
22. Shaw. *The Oxford Ancient History of Egypt*, p. 271.
23. Potter. *The Story of Religion*, p. 15.
24. Hall. *The Secret Destiny of America*, p. 24-27
25. Osman. *Moses and Akhenaten*, p. 63
26. Shaw. *The Oxford Ancient History of Egypt*, p. 272.
27. Ibid, p. 281.
28. Carson. "Hidden Black Kings and Queens of Ancient Egypt with Billy Carson" 42:40.
29. Shaw. *The Oxford Ancient History of Egypt*, p. 282.
30. Ibid, p. 285.
31. Breasted. *Ancient Records of Egypt: Vol III*, 11.

11. The Biblical Patriarchs: Abraham to Joseph

1. *Chapter Eleven*
   Maxwell. *That Old-Time Religion*, p. 75-76.
2. Osman. *Moses and Akhenaten*, p. 13
3. Osman. *The Egyptian Origins of King David and the Temple of Solomon*, p. 45.

Biblical Name

1. Ibid, p. 1.
2. Osman. *The Egyptian Origins of King David and the Temple of Solomon*, p. 43-44.
3. Polano. *Selections from the Talmud*, p. 34-38.
4. Ibid, p. 30.
5. Ibid, p. 30-42.
6. Osman. *The Egyptian Origins of King David and the Temple of Solomon*, p. 44.
7. Ibid, p. 46.
8. Ibid, p. 50.
9. Strom. "Cutting Truth About Circumcision."
10. Josephus. *Against Apion I*, 167-170.
11. Osman. *The Egyptian Origins of King David and the Temple of Solomon*, p. 49.

12. Biglino. *Gods of the Bible*, p. 203.
13. Sitchin. *The Wars of Gods and Men*, p. 290-91.
14. Ibid, 294-96.
15. Josephus. *The Antiquities of the Jews*, 1.59.
16. Sitchin. *The Wars of Gods and Men*, 290.
17. Ibid, p. 193, 300-301.
18. Osman. *The Egyptian Origins of King David and the Temple of Solomon*, p. 74.
19. Polano. *Selections from the Talmud*, p. 51.
20. Ibid, p. 83-85.
21. Osman. *The Egyptian Origins of King David and the Temple of Solomon*, p. 76.
22. O'Brien. *Genius of the Few*, p. 27.
23. Osman. *The Egyptian Origins of King David and the Temple of Solomon*, p. 25.
24. Polano. *Selections from the Talmud*, p. 84.
25. Osman. *The Egyptian Origins of King David and the Temple of Solomon*, p. 26-28.
26. Ibid, p. 6-7.

12. MOSES: BIBLICAL PATRIARCH OR EGYPTIAN PHARAOH?

1. Osman, *The Egyptian Origins of King David and the Temple of Solomon*, p. 89.
2. *Josephus*, Ant II 15:2.
3. Polano. *Selections from the Talmud*, p. 141.
4. Ibid, p. 98.
5. Ibid, p. 104
6. Osman, *Moses and Akhenaten*, p. 217-220
7. Ibid, p. 105
8. Exodus 1:15-16
9. Osman, *Moses and Akhenaten*, p. 14
10. Exodus 1:17-22
11. Polano. *Selections from the Talmud*, p. 108
12. Exodus 6:14-25
13. Josephus. *Against Apion I,* 230-266
14. Exodus 2:1-10
15. Polano, *Selections from the Talmud*, p. 107
16. Shaw. *The Oxford History of Ancient Egypt*, p. 269.
17. Osman, *Egyptian Origins*, p. 6
18. Polano. *Selections from the Talmud*, p. 110
19. Ibid, p. 108
20. Ibid, p. 112-115
21. Osman, *Moses and Akhenaten*, p. 23. Polano. *Selections from the Talmud*, p. 86.
22. Ibid, p. 7.
23. Biglino, *Gods of the Bible*, p. 134
24. O'Brien. *The Genius of the Few*, p. 29.
25. Polano. *Selections from the Talmud*, p. 116
26. Exodus 7:9-12
27. Sitchin. *Divine Encounters*, p. 348.
28. Osman, *Moses and Akhenaten*, p. 176-179

29. Shaw. *The Oxford Ancient History of Egypt*, p. 276.
30. Ibid, p. 174
31. Ibid, p. 179
32. *Brown Drivers Briggs Hebrew and English Lexicon*, p. 639a
33. Smith. *The Chaldean Account of Genesis*, p. 183-84.
34. Finkelstein. *David and Solomon*, p. 223.
35. Pike. *Morals and Dogma*, p. 232.

13. MYSTERIOUS ASPECTS OF YAHWEH WITH MOSES

1. *Chapter Thirteen*
   Madden Jones. *The Yahweh Encounters*, p. 5.
2. Biglino. *Gods of the Bible*, p. 106.
3. O'Brien. *The Shining Ones*, p. 167.
4. Ibid, p. 9.
5. Biglino. *Gods of the Bible*, p. 63.
6. Madden Jones. *The Yahweh Encounters*, p. 10.
7. Madden Jones. *The Yahweh Encounters*, p. 10.
8. Ibid, p. 9-12.
9. Madden Jones. *The Yahweh Encounters*, p. 13.
10. Ibid, p. 82-84.
11. O'Brien. *The Genius of the Few*, p. 186, 207-8.
12. Osman. *The Egyptian Origins of King David and the Temple of Solomon*, p. 140.
13. Biglino. *Gods of the Bible*, p. 268.
14. Ibid, p. 267.
15. Ibid, p. 268.
16. Madden Jones, *The Yahweh Encounters*, p. 79.
17. Clark. *Etymological Dictionary of Biblical Hebrew*, p. 120.
18. Biglino. *Gods of the Bible*, p. 281.
19. Ibid p. 271.
20. Madden Jones. *The Yahweh Encounters*, p. 123.
21. Ibid, p. 76.
22. Ibid, p. 84.
23. Ibid, p. 103.
24. Ibid, p. 111.
25. Ibid, p. 85.
26. Haran. *Temples and Temple Service in Ancient Israel*, p. 176.
27. Dunn. *The Giza Power Plant*.
28. Biglino. *Gods of the Bible*, p. 276.
29. Madden Jones. *The Yahweh Encounters*, p. 88.
30. Josephus. *Antiquities of the Jews*, 3.7.1-5.
31. Madden Jones. *The Yahweh Encounters*, p 89.
32. Biglino. *The Gods of the Bible*, p. 275.
33. Josephus. *Antiquities of the Jews*, 3.7.5.
34. Madden Jones. *The Yahweh Encounters*, p. 90-91.
35. O'Brien. *The Genius of the Few*, p. 206.

36. Josephus. *Antiquities of the Jews,* 3.8.9.
37. Biglino. *Gods of the Bible*, p. 277-278.

14. Giants, Nephilim, Watchers, or Anunnaki Hybrids? All of the Above

1. *Chapter Fourteen*
   The Book of Giants, p. 2.
2. Gaster. *Myth, Legend, and Custom in the Old Testament*, p. 79.
3. Sitchin. *The Stairway to Heaven*, p. 193.
4. Sitchin. *The Wars of Gods and Men*, p. 197.
5. *Epic of Gilgamesh,* Tablet V v, 5-9, p. 39.
6. Ibid. p, 241.
7. O'Brien, *The Shining Ones*, p. 45.
8. Ibid, p. 119
9. Sitchin. *The End of Days*, p. 53-54.
10. Sitchin. *The Lost Book of Enki*, p. 205.
11. Sitchin. *The Wars of Gods and Men*, p. 196-97.
12. O'Brien. *The Genius of the Few*, p. 73.
13. *The Book of the Secrets of Enoch*, Book I v 5, p. 2.
14. O'Brien. *The Genius of the Few*, p. 73.
15. O'Brien. *The Genius of the Few*, p. 93.
16. *The Book of Giants,* p, 14.
17. Ibid, p. 14-20.
18. Ibid, p. 3-4.
19. Biglino. *Gods of the Bible*, p. 125.
20. Josephus. *The Antiquities of the Jews*. 1.72-74.
21. *The Book of Giants*, p. 18.
22. Sitchin. *The Lost Book of Enki*, p. 274.
23. Biglino. *Gods of the Bible*, p. 121.
24. Ibid, p. 126-127.

15. King David

1. *Chapter Fifteen*
   Osman. *Egyptian Origins of King David and the Temple of Solomon*, p. 1.
2. Shaw, *The Oxford History of Ancient Egypt*, p. 299.
3. Osman, *Egyptian Origins of King David and the Temple of Solomon*, p. 15.
4. Ibid, p. 18.
5. Garsiel. "The Book of Samuel: Its Composition, Structure and Significance as a Historiographical Source" p. 4-7.
6. Ibid, p. 6.
7. Ibid, p. 5.
8. Osman, *The Egyptian Origins of King David and the Temple of Solomon*, p. 10.
9. Davies. "'House of David' Built on Sand: the Sins of the Biblical Maximizers.
10. Finkelstein. *David and Solomon*, p. 130.
11. Osman, *The Egyptian Origins of King David and the Temple of Solomon*, p. 12-13.

12. Finkelstein. *David and Solomon*, p. 269.
13. Ibid, p. 274.
14. Gardiner. *The Egyptians*, p. 198.
15. Shaw. *The Oxford History of Ancient Egypt,* p, 256.
16. Osman, *The Egyptian Origins of King David and the Temple of Solomon*, p. 27.
17. Ibid, p. 31.
18. Bullock. *The Story of Sinuhe,* p. 37-44.
19. Ibid, p. 47.
20. Ibid, p. 47.
21. Osman, *The Egyptian Origins of King David and the Temple of Solomon*, p. 71.
22. Williams, *Literature of Ancient Egypt,* p. 59.
23. Breasted, *Ancient Records of Egypt Vol 2*, 139-141.
24. Osman, *The Egyptian Origins of King David and the Temple of Solomon*, p. 41.
25. *Encyclopaedia Judaica*, vol 6, p. 315.
26. Selman. *1 Chronicles: An Introduction and Commentary*, p. 199.
27. Osman. *The Egyptian Origins of King David and the Temple of Solomon*, p. 41.
28. Good. *Irony in the Old Testament,* p. 35-37.
29. Osman. *The Egyptian Origins of King David and the Temple of Solomon*, p. 61-62.

16. THE REAL SOLOMON?

1. *Chapter Sixteen*
   Osman. *The Egyptian Origins of King David and the Temple of Solomon*, p. 102.
2. Handcock. *Selections from the Tell El-Amarna Letters*, p. 4-12.
3. Lichtheim. *Ancient Egyptian Literature Volume II: The New Kingdom*, p. 77.
4. Shaw. *The Oxford History of Ancient Egypt*, p. 218.
5. Moran. *The Amarna Letters,* p. 8.
6. Ash. *David, Solomon, and Egypt: A Reassessment*, p. 115.
7. Osman. *The Egyptian Origins of King David and the Temple of Solomon*, p. 107.
8. Ibid, p. 113.
9. Ibid, p. 109.
10. Ibid, p. 109-110.
11. Ibid, p. 97-99.
12. Finkelstein. *David and Solomon*, p. 161.
13. Ibid, p. 130-36.
14. Flemming. *The History of Tyre*, p. 8-9.
15. Finkelstein. *David and Solomon*, p. 153.
16. Lemche. "The Origin of the Israelite State," p. 45.
17. Finkelstein. *David and Solomon*, p. 162, 172.
18. Pike. *Morals and Dogma*, p. 232.
19. Finkelstein. *David and Solomon*, p. 203.
20. Ibid, p. 221.
21. Hall. *The Secret Teachings of All Ages*, p. 512-513.
22. Ibid, p. 95.
23. Finkelstein, *David and Solomon*, p. 162.
24. Pike. *Morals and Dogma*, p. 195, 210, 238. Hall. *Secret Teachings of All Ages*, p. 512-13.

25. "Testament of Truth", *Nag Hammadi Scriptures*, p. 626-627.
26. Josephus. *Complete Works of Josephus*, 8.42.
27. Ebeling. *The Secret History of Hermes Trismegistus*, p. 28.
28. Josephus. *Complete Works of Josephus*, 8.45-47.
29. *Testament of Solomon*, p. 49.
30. Ebeling. *The Secret History of Hermes Trismegistus*, p. 27.
31. Ibid, p. 27.
32. Ibid, p. 39-42.
33. Blavatsky. *The Secret Doctrine: Volume 2*, p. 154.
34. Madden Jones. *The Yahweh Encounters*, p. 208.
35. Finkelstein. *David and Solomon*, p. 222-23.
36. Bramley. *The Gods of Eden*, p. 89.

17. Solomon's Temple is Not What You Think

1. *Chapter Seventeen*
   Osman, *The Egyptian Origins of King David and the Temple of Solomon*, p. 166.
2. Josephus. *The Antiquities of the Jews*, 8.69.
3. Clarke, *Adam Clarke's Commentary on the Holy Bible*, p. 396.
4. Breasted. *Ancient Records of Egypt: Vol II*, 890.
5. Biglino. *Gods of the Bible*, p. 63.
6. Osman, *The Egyptian Origins of King David and the Temple of Solomon*, p. 140, 150.
7. Hall. *Secret Teachings of All Ages*, p. 247.
8. Mackey. *Encyclopedia of Freemasonry and Kindred Sciences Vol II*, p. 797.
9. Madden Jones, *The Yahweh Encounters*, p. 223-225.
10. Biglino. *Gods of the Bible*, p. 250.
11. Ibid, p. 282.
12. Madden Jones, *The Yahweh Encounters*, p. 214.
13. Clarke. *Adam Clarke's Commentary on the Holy Bible*, p. 341.
14. Osman. *The Egyptian Origins of King David and the Temple of Solomon*, p. 150-51.
15. Ibid, p. 157.
16. Mackey. *Encyclopedia of Freemasonry and Kindred Sciences Vol II*, p. 567-68.
17. Madden Jones. *The Yahweh Encounters*, p. 216.
18. Mackey. *Encyclopedia of Freemasonry and Kindred Sciences Vol II*, p. 568.
19. Pike. *Morals and Dogma*, p. 12.
20. Mackey. *Encyclopedia of Freemasonry and Kindred Sciences Vol II*, p. 567-68.
21. Madden Jones. *The Yahweh Encounters*, p. 216-17.
22. Hall. *Secret Teachings of All Ages*, p. 248-49.
23. Madden Jones. *The Yahweh Encounters*, p. 216
24. Mackey. *Encyclopedia of Freemasonry and Kindred Sciences Vol II*, p. 566
25. Waite. *A New Encyclopedia of Freemasonry*, p. 280.
26. Pike. *Morals and Dogma*, p. 12.
27. Madden Jones. *The Yahweh Encounters*, p. 218-19
28. Ibid, p. 79.
29. Madden Jones. *The Yahweh Encounters*, p. 221-22.
30. Ibid, p. 222.

31. Pike. *Morals and Dogma*, p. 196.
32. Madden Jones. *The Yahweh Encounters*, p. 216-229.
33. Josephus. *The Antiquities of the Jews*, 8.76.

18. The Strange Enigma of Hiram

1. *Chapter Eighteen*
   Josephus. *The Antiquities of the Jews* 8.76
2. Mackey. *Encyclopedia of Freemasonry and Its Kindred Sciences Vol. I*, p. 329.
3. Pike. *Morals and Dogma*, p. 77.
4. Josephus. *Antiquities of the Jews* 8.144-149.
5. Pike. *Morals and Dogma*, p. 76-77.
6. Maroke. *The Phoenicians*, p. 168-169.
7. Mackey. *Encyclopedia of Freemasonry and Its Kindred Sciences Vol. I*, p. 330.
8. Washbourne. *A Freemason's Pocket Companion*, p. 20.
9. Hall. *The Secret Teachings of All Ages*, p. 187.
10. Mackey. *Encyclopedia of Freemasonry and Its Kindred Sciences Vol. I*, p. 330.
11. Ibid, p. 4.
12. Pike. *Morals and Dogma*, p. 79.
13. Mackey. *Encyclopedia of Freemasonry and Its Kindred Sciences Vol. I*, p. 4.
14. Łajtar. "Deir El-Bahari in the Hellenistic and Roman Periods", p. 13.
15. Breasted. *Ancient Records of Egypt Vol II*, 912.
16. Łajtar. "Deir El-Bahari in the Hellenistic and Roman Periods", p. 13-14.
17. Ibid, p. 14-15.
18. Ibid, p. 25.
19. Josephus. *Against Apion I* 1.26-28.
20. Ibid, p. 26, 28.
21. Wilchen. *Festschrift für Gorg Ebers*, p. 144-45.
22. Pike. *Morals and Dogma*, p. 77.
23. Ebeling. *The Secret History of Hermes Trismegistus*, p. 27.
24. Benner. *Ancient Hebrew Dictionary*, p. 58.
25. Ibid, p.
26. Hall. *Secret Teachings of All Ages*, p. 191.
27. Ibid, p. 454.
28. Hall. *The Secret Teachings of All Ages*, p. 190-193.
29. Pike. *Morals and Dogma*, p. 210.

19. Elijah, Ezekiel, Chariots, and Sky Armies

1. *Chapter Nineteen*
   Blumrich. *Spaceships of Ezekiel*, p. 10.
2. Biglino. *Gods of the Bible*, p. 62.
3. Blumrich. *Spaceships of Ezekiel*, p. 5.
4. Ibid, p. 14.
5. Ibid, p. 30-35.
6. Ibid, p. 31-41.

7. Pike. *Morals and Dogma*, p. 239-240.
8. Sitchin. *The Stairway to Heaven*, p. 193.
9. Sitchin. *The Stairway to Heaven*, p. 201.
10. Sitchin. *The Stairway to Heaven*, p. 201.
11. Blumrich. *Spaceships of Ezekiel*, p. 57-60.
12. Ibid, p. 64.
13. Anderson. "Structures Technology."
14. Blumrich. *Spaceships of Ezekiel*, p. 25-31.
15. Ibid, p. 67.
16. *The Book of the Secrets of Enoch*, I v.5, p. 2.
17. *The Book of the Secrets of Enoch*, XXIX v.3, p. 36.
18. Benner. *The Ancient Hebrew Language and Alphabet*, p. 151.
19. Ibid, p. 136.
20. Ibid, p. 176.
21. Ibid, p. 184.
22. Ibid, p. 179.
23. Ibid, p. 169.
24. Livy. *The History of Rome*, p. 762.
25. Josephus. *Wars of the Jews VI*, 5.289, 297-299.
26. Plutarch. *Parallel Lives Vol. II*, p. 301.

20. Discovering Yahweh: The Bible and Stories of the Anunnaki

1. *Chapter Twenty*
   Sitchin. *The 12th Planet*, p. 388.
2. Hoffner. *Hurrian Myths*, p. 40-65.
3. Sitchin. *The Wars of Gods and Men*, p. 76-86.
4. Ibid, p. 345-350 for timetable until 2023 BCE. Sitchin. *The End of Days*, p. 274-75 for 2160 BCE to 0.
5. Goetze. "Kingship in Heaven" *Ancient Near Eastern Texts*, p. 120-28.
6. Sitchin. *The 12th Planet*, p. 347-353.
7. Ibid, p. 337.
8. Dalley. *Myths from Mesopotamia*, p. 17.
9. Ibid, p. 17-18.
10. Sitchin. *The 12th Planet*, p. 357.
11. Sitchin. *The Lost Realm*, p. 253-54.
12. Meyer. *Nag Hammadi*, p. 347-48.
13. Sitchin. *The 12th Planet*, p. 359.
14. Kramer. *Sumerian Mythology*, p. 74-76.
15. Malan. *The First and Second Books of Adam and Eve*, p. 8-66.
16. Ibid, p. 66-70.
17. Sitchin. *The Lost Realm*, p. 30, 41-42.
18. Meyer. *Nag Hammadi*, p. 526.
19. Sitchin. *Divine Encounters*, p. 43-47.
20. Ibid, p. 84.
21. Manetho. *Aegyptiaca*, p. 3-7.

22. Hapgood. *The Earth's Shifting Crust*, p. 277.
23. Ibid, p. 294.
24. Ibid, p. 269-270.
25. Dalley. *Myths of Mesopotamia*, p. 18-29.
26. Sitchin. *Lost Realms*, p. 255.
27. Ibid, p. 28-31
28. *Epic of Gilgamesh*, XI 1-70.
29. Smith. *The Chaldean Account of Genesis*, p. 164.
30. Sitchin. *The 12th Planet*, p. 397.
31. Dalley. *Myths of Mesopotamia*, p. 32-33.
32. *The Epic of Gilgamesh* XI 157-206.
33. Dalley. *Myths of Mesopotamia*, p. 290-92.
34. Sitchin. *The Wars of Gods and Men*, p. 129-130.
35. Sitchin. *The 12th Planet*, p. 89-127.
36. Sitchin. *The Wars of Gods and Men*, p. 38-41.
37. Murray. *Ancient Egyptian Legends*, p. 59-73.
38. Sitchin. *The Wars of Gods and Men*, p. 155-56.
39. Ibid, p. 127-28.
40. Mirelam. "A New Manuscript of Lugal-E, Tablet IV," p. 4-6.
41. Sitchin. *The Wars of Gods and Men*, p. 162-64.
42. Ibid, p. 165-67, 173-75.
43. Ibid, 169-171.
44. Ibid, p. 174-184.
45. Ibid, p. 192-97.
46. George. *The Tamarisk, the Date-Palm, and the King*, p. 75-90.
47. Jacobsen. *The Sumerian Kings List*, p. 85-93.
48. Boscawen. *Transactions of the Society of Biblical Archaeology Vol V*, p. 304-311.
49. Sitchin. *The Wars of Gods and Men*, p. 216-228.
50. Sitchin. *The Lost Realms*, p. 272-73.
51. *Manetho*, p. 7, 17-21.
52. Wallis Budge. *History of Ethiopia, Nubia, and Abyssinia*, p. xiv-xvi, 129.
53. Kramer. "Inanna and Enki: The Transfer of the Arts of Civilization," *Myths of Enki*, p. 58-68.
54. Sitchin. *The Wars of Gods and Men*, p. 229-250.
55. Kramer. "Enmerker and the Lord of Aratta," *History Begins at Sumer*, p. 22-27.
56. Bramley. *The Gods of Eden*, p. 66-67.
57. Cooper. *The Curse of Agade*, p. 50-63.
58. Sitchin. *The Wars of Gods and Men*, p. 270-77.
59. Ibid, p. 277-280.
60. Ibid, p. 290-307.
61. Ibid, p. 325-338.
62. Michalowski. *The Lamentation over the Destruction of Sumer and Ur*, p.
63. Sitchin. *The Cosmic Code*, p. 194-197.
64. Sitchin. *The End of Days*, p. 137.
65. Spencer. *Rosicrucian Questions and Answers*, p. 50, 59.
66. Moran. *The El Amarna Letters*, p. xxiv-9.
67. Sitchin. *The End of Days*, p. 171, 188.
68. Ibid, p. 252.

69. Ibid, p. 191-210.
70. Sitchin. *The Cosmic Code*, p. 275.
71. Sitchin. *The End of Days*, p. 219-223.
72. Sitchin. *The Wars of Gods and Men*, p. 20-22.
73. Hall. *The Secret Teachings of All Ages*, p. 10.
74. O'Brien. *The Genius of the Few*, p. 29-30.
75. Sitchin. *Divine Encounters*, p. 378.
76. Ibid, p. 347-379.
77. Fink. "The Unrecognizability of God in Balangs and Xenophanes"

21. Discovering the Creation of Jesus and the Church – The Mysteries from Antiquity to Constantine the Great

1. *Chapter Twenty-One*
   Pike. *Morals and Dogma*, p. 76-77.
2. Bramley. *The Gods of Eden*, p. 113.
3. Maxwell. *That Old-Time Religion*, p. 36-53.
4. Sri Yukteswar. *The Holy Science*, p. xiv-xviii.
5. Heckethorn. *Secret Societies of All Ages and Countries*, p. 3.
6. Hall. *The Secret Destiny of America*, p. 53-54.
7. Lewis. *Rosicrucian Questions and Answers*, p. 17-21.
8. Ibid, p. 112-113.
9. Ibid, p. 33-51.
10. Moran. *The Amarna Letters*, p. 1-20.
11. Bramley. *The Gods of Eden*, p. 94-102.
12. Heckethorn. *Secret Societies of All Ages and Countries*, p. 34-38.
13. Miller. *Occult Theocrasy*, p. 65-66.
14. Hall. *The Secret Teachings of All Ages*, p. 10-11.
15. Heckethorn. *Secret Societies of All Ages and Countries*, p. 42-50.
16. Ibid, p. 11-12.
17. Icke. *The Biggest Secret*, p. 96, 110. Miller. *Occult Theocrasy*, p. 90-91.
18. Lewis. *Rosicrucian Questions and Answers*, p. 88.
19. Miller. *Occult Theocrasy*, p. 108-111.
20. Ibid, p. 163-64.
21. Heckethorn. *Secret Societies of All Ages and Countries*, p. 142.
    Chapter Twenty-Two

22. Creation of the Christian Church and Its Values

1. Charles River Editors. *The Ancient World's Most Mysterious Religious Cult*.
2. Vermaseren. *Mithras, the Secret God*, p. 59.
3. Leedom. *The Book Your Church Doesn't Want You to Read Book II*, p. 275-81.
4. Ibid, p. 282.
5. Graves. *The World's Sixteen Crucified Saviors*, p. 12, 58.
6. Massey. *The Natural Genesis Vol II*, p. 438.

7. Leedom. *The Book Your Church Doesn't Want You to Read Book II*, p. 29.
8. Bramley. *The Gods of Eden*, p. 78.
9. Lewis. *Rosicrucian Questions and Answers*, p. 89.
10. Icke. *The Biggest Secret*, p. 123-4.
11. Ibid, p. 148-150.
12. Koestler. *The Thirteenth Tribe*, p. 17.
13. Ibid, p. 15.
14. Ibid, p. 80.
15. Lewis. *Rosicrucian Questions and Answers*, p. 108.
16. Leedom. *The Book Your Church Doesn't Want You to Read Book II*, 283-88.
17. Icke. *The Biggest Secret*, p. 144.
18. Creighton. *The English Historical Review*, p. 670.
19. Miller. *Occult Theocrasy*, p. 140-142.
20. Bramley. *The Gods of Eden*, p.159-62.
21. Baigent. *Holy Blood, Holy Grail*, p. 106.
22. Ibid, p. 206.
23. Ibid, p. 141.
24. Icke. *The Biggest Secret*, p. 167.
25. Marino. *Christopher Columbus, the Last Templar*, p. 30, 49.
26. Bramley. *The Gods of Eden*, p. 155.
27. Heckethorn. *The Secret Societies of All Ages and Countries*, p. 159.
28. Miller. *Occult Theocrasy*, p. 143-45.
29. Bramley. *The Gods of Eden*, p. 157-58.
30. Ibid, p. 158, 248-9.
31. Icke. *The Biggest Secret*, p. 154.
32. Bramley. *The Gods of Eden*, p. 197-200.

Chapter Twenty-Three

23. LIES AND THE AGE OF DISCOVERIES

1. Maurino. *Christopher Columbus, the Last Templar*, p. 8.
2. Lewis. *Rosicrucian Questions and Answers*, p. 53.
3. Hall. *America's Assignment with Destiny*, p. 187.
4. Marino. *Christopher Columbus, the Last Templar*, p. 35.
5. Hall. *Secret Destiny of America*, p. 62.
6. Hall. *America's Assignment with Destiny*, p. 186-92.
7. Ibid, p. 194.
8. Icke. *The Biggest Secret*, p. 124.
9. Hall. *America's Assignment with Destiny*, p. 192.
10. Ibid, p. 149-154.
11. Sertima. *They Came Before Columbus*, p. 74.
12. Hall. *America's Assignment with Destiny*, p. 152.
13. Hall. *Secret Destiny of America*, p. 90.
14. Maurino. *Christopher Columbus, the Last Templar*, p. 17.
15. Hall. *America's Assignment with Destiny*, p. 166.
16. Ibid, p. 151.
17. Sitchin. *The Lost Realms*, p. 9.

18. Ibid, p. 10-16.
19. Lewis. *Rosicrucian Questions and Answers*, p. 128.
20. Bramley. *The Gods of Eden*, p. 201-5.
21. Icke. *The Biggest Secret*, p. 161.
22. Bramley. *The Gods of Eden*, p. 84-88.
23. Chibuife. *Dahomey and the Slave Trade*, p. 16.
24. Bramley. *The Gods of Eden*, p. 221-26.
25. Icke. *The Biggest Secret*, p. 170.
26. Cusack. *The Black Pope*, p. 42-43.
27. Ibid, p. 24.
28. Ibid, p. 60.
29. Ibid, p. 16.
30. Roberts. *The Mythology of Secret Societies*, p. 118-23.
31. Hall. *The Secret Teachings of All Ages*, p. 480-84.
32. Hall. *America's Assignment with Destiny*, p. 195-6.
33. Hall. *The Secret Destiny of America*, p. 91.
34. Icke. *The Biggest Secret*, p. 162.
35. Hall. *The Secret Destiny of America*, p. 57.
36. Webster. *Secret Societies and Subversive Movements*, p. 120.
37. Icke. *The Biggest Secret*, p. 126.
38. Bramley. *The Gods of Eden*, p. 226-31.
39. Icke. *The Biggest Secret*, p. 129.
40. Lewis. *Rosicrucian Questions and Answers*, p. 17, 175-77.

24. The New, Shiny England and Move Towards Domination

1. *Chapter Twenty-Four*
   Bramley. *The Gods of Eden*, p. *233-35.*
2. Icke. *The Biggest Lie*, 183-84.
3. Bramley. *The Gods of Eden*, p. 232.
4. Icke. *The Biggest Lie*, p. 210.
5. Miller. *Occult Theocrasy*, p. 183.
6. Roberts. *The Mythology of Secret Societies*, p. 103-105.
7. Ibid, p. 209.
8. Bramley. *The Gods of Eden*, p. 299.
9. Lewis. *Rosicrucian Questions and Answers*, p. 77.
10. Roberts. *The Mythology of Secret Societies*, p. 118-126.
11. Miller. *Occult Theocrasy*, p. 183-87.
12. Ibid, p. 275-8.
13. Ibid, p. 279.
14. Hall. *The Secret Destiny of America*, p. 94-5.
15. Bramley. *The Gods of Eden*, p. 179-80.
16. Ibid, p. 288-93.
17. Harvey. *Through the Storm, Through the Night*, p. 6.
18. Ibid, p. 16-19.
19. Chibuife. *Dahomey and the Slave Trade*, p. 23-25.

20. Bramley. *The Gods of Eden,* p. 294-5.
21. Icke. *The Biggest Lit,* p. 224.
22. Bramley. *The Gods of Eden,* p. 326-27, 353-54.
23. Icke. *The Biggest Lie,* p. 149, 241, 306.
24. Lewis. *Rosicrucian Questions and Answers,* p. 138-41.
25. Bramley. *The Gods of Eden,* p. 328-34.

Chapter Twenty-Five

25. Project Domination

1. Kasten. *Dark Fleet.*
2. Jones. *The Red Tails.*
3. Friedman. *Catpured.*
4. Kasten. *Secret Journey to Planet Serpo.*
5. Kastern. *Alien World Order.*
6. Webre. *The Omniverse.*

Image Credits

Figure 1: Moai Monoliths *Credit: RIVI CC BY SA 2.0 Wiki Commons*Figure 3: Nevali Cori Statues *Credit: Dosseman, CC BY-SA 2.0, Wiki Commons*Figure 2: Urfa Man*Credit: Dosseman CC BY-SA 4.0, Wiki Commons*Figure 4: Petroglyphs at Chaco Canyon *Credit: Fran Hogan CC: Public Domain*Figure 5: Basalt Olmec Head *Credit Gary Todd, CC BY-SA 2.0, Wiki Commons*Figure 6: Paracas Skulls *Credit: Marcin Tlustochowia CC: BY-SA 2.0, Wiki Commons*Figure 7: Sagittal Suture *Credit: Public Domain, Wiki Commons* Adjustments and enhancements made for clarityFigure 8: Carin T Petroglyphs *Credit: Voxlivia CC: BY-SA 2.0, Wiki Commons* (Photo is enhanced to see the formation of petroglyphs)Figure 9: Hopi Mesas and Orion Correlation. *Credit: NorthstarFigure 10: Oannes Credit: Dr. Regosistvan CC BY-SA 4.0, Wiki Commons*Figure 11: Above: Apkallu holding flowers in their hand with watches around both wrists. *Credit: Shukit Muhammed Amin CC BY-SA 4.0, Wiki Commons*Left: Apkallu holding the handbag *Credit: Dr. Osama Shukit Muhammed Amin CC BY-SA 4.0, Wiki Commons*(Pictures adjusted for clarity)Table 1: Names of Biblical characters correlated with names from *The Lost Book of Enki*Table 2: Table of Patriarchs Across MesopotamiaTable 3: Ugaritic PantheonTable 4: Chronology of the 18th DynastyFigure 12: Amenhotep III *Credit:*

*Public Domain Wiki Commons*Figure 14: Akhenaten *Credit: Public Domain*Figure 15: Queen Nefertiti *Credit: Miguel Hermoso Cuesta. CC: BY-SA 4.0 Wiki Commons*Figure 16: Meritaten Eldest Daughter of Akhenaten and Nefertiti *Credit: Miguel Hermoso Cuesta CC: BY-SA 3.0, Wiki Commons*Figure 17: Neferneferuaten Tasherit and Neferneferure. Daughters of Akhenaten and Nefertiti*Credit: Public Domain, Wiki Commons**Table 5: Table of the Pharaohs and Patriarch*Figure 18: Cherubim*Credit: Recreation from M. Biglino.*Figure 19: Ark with Cherubim Positioning*Credit: Recreation from M. Biglino.*Figure 20: Solomon Ring *Credit: Pblpitt CC: By-SA 4.0*Figure 22: (Top) Artist interpretation of single cherubim with dimensions. (Below) Artist interpretation of two cherubim touching wing to wing in the Holy of Holies *Artist Rendition © 2024 Jennifer Lahr*Figure 23: Artist interpretation of Pillarprovides a size perspective with an average human height of a 5'6" male.*Artist Rendition © 2024 Jennifer Lahr*Figure 24: Artist's interpretation of single bronze basin. Provided: Size perspective with an average human height of a 5'6" male*Artist Rendition © 2024 Jennifer Lahr*Figure 25: Above: Brazen Sea. Artist Interpretation from Jewish Encyclopedia, 1906. *Credit: Public Domain, Wiki Commons.*Right: CSIRO Earth Receiving Station Satellite Dish Droughty Point Hobart Tasmania.*Credit: Bruce Miller CC BY-SA 4.0, Wiki Commons*Below: Artist Interpretation of Molten Sea.Provided: Size perspective with an average human height of a 5'6" male.*Artist Rendition © 2024 Jennifer Lahr*Figure 26: Artist Interpretation of Solomon's Temple Complex-Provided: Size perspective given with three average-height males*Artist Rendition © 2024 Jennifer Lahr*

www.ingramcontent.com/pod-product-compliance
Lightning Source LLC
Chambersburg PA
CBHW060834280326
41934CB00007B/776

* 9 7 9 8 9 8 7 1 2 2 4 9 5 *